Joseph Ray

Ray's arithmetic: Second book

Intellectual arithmetic by induction and analysis

Joseph Ray

Ray's arithmetic: Second book
Intellectual arithmetic by induction and analysis

ISBN/EAN: 9783337278076

Printed in Europe, USA, Canada, Australia, Japan

Cover: Foto ©berggeist007 / pixelio.de

More available books at **www.hansebooks.com**

ECLECTIC EDUCATIONAL SERIES.

RAY'S ARITHMETIC, SECOND BOOK.

INTELLECTUAL
ARITHMETIC,

BY

INDUCTION AND ANALYSIS.

By JOSEPH RAY, M. D.,
LATE PROFESSOR OF MATHEMATICS IN WOODWARD COLLEGE.

ONE THOUSANDTH EDITION—IMPROVED.

CINCINNATI:
WILSON, HINKLE & CO.
PHIL'A: CLAXTON, REMSEN & HAFFELFINGER.
NEW YORK: CLARK & MAYNARD.

RAY'S MATHEMATICAL WORKS.

TYPE ENLARGED—NEW ELECTROTYPE PLATES.

Each Book *of Ray's Arithmetical Course, also of the Algebraic, is a complete work in itself, and is sold separately.*

FIRST BOOK.
PRIMARY LESSONS AND TABLES; simple and progressive Mental Lessons, for little learners.

SECOND BOOK.
INTELLECTUAL ARITHMETIC, by Induction and Analysis; a thorough course on Intellectual Arithmetic.

THIRD BOOK.
PRACTICAL ARITHMETIC, by Induction and Analysis; a simple, thorough work for schools and private students.

KEY TO RAY'S ARITHMETIC, THIRD BOOK.

RAY'S HIGHER ARITHMETIC. Principles of Arithmetic, analyzed and applied. For advanced students and business men.

KEY TO RAY'S HIGHER ARITHMETIC.

ELEMENTARY ALGEBRA.
RAY'S ALGEBRA, FIRST BOOK, for Common Schools and Academies; a simple, progressive, elementary treatise.

HIGHER ALGEBRA.
RAY'S ALGEBRA, SECOND BOOK, for Academies, and for Colleges; a progressive, lucid, and comprehensive work.

KEY TO RAY'S ALGEBRA, FIRST AND SECOND BOOKS.

Entered according to Act of Congress, in the year Eighteen Hundred and Fifty-Seven, by WINTHROP B. SMITH, in the Clerk's Office, of the District Court of the United States, for the Southern District of Ohio.

Entered according to Act of Congress, in the year Eighteen Hundred and Sixty, by W. B. SMITH, in the Clerk's Office of the District Court of the United States, for the Southern District of Ohio.

PREFACE.

Few works on Intellectual Arithmetic have received more unqualified approbation, and a more extensive patronage, than this, which, for several years, has been published under the title—Ray's Arithmetic, Part Second.

The numerous editions demanded, have again rendered necessary a renewal of the plates, which has afforded an opportunity for REMODELING the work.

Many important improvements have been introduced, with a design to impart completeness, and give a concise and progressive course of arithmetical analysis.

The volume in its present form, embraces,

1st. Exercises on the primary principles and their applications, interspersed with appropriate models of analysis and frequent reviews.

2d. A progressive and comprehensive presentation of Fractions, intended to render the subject intelligible and attractive to the pupil.

3d. A General Review, designed to test the pupil's knowledge of principles, preparatory to the applications of mental analysis which follow.

4th. Percentage, Gain and Loss, Interest, and their applications.

The value of Intellectual Arithmetic is so highly appreciated by instructors, that little need be said in its commendation.

When properly taught, it is one of the most useful and interesting studies in which pupils can engage, and should be omitted by no one.

By its study, learners are taught to reason, to analyze, to think for themselves; while it imparts confidence in their own reasoning powers, and strengthens the mental faculties.

HINTS TO TEACHERS.

LET the pupils be classified with reference to their attainments and abilities. The recitation should be short and spirited, every pupil being required to give undivided attention to the question before the class.

Generally, while reciting, the pupils should be permitted to have their books open before them—the test of having properly studied the lesson, being the *readiness* and *accuracy* with which the several questions are analyzed and answered.

The explanations and operations termed ANALYSIS, are intended as *Model Solutions*, pointing out to the learner the *manner* in which the questions in the lessons are to be solved and explained.

The pupil should be required to furnish a similar explanation to each of the succeeding questions, and to give, not only a correct answer, but also, the *reason* for the method by which he obtained it.

A method of solving questions in Mental Arithmetic, now much used, is the following, called the "*Four Step Method.*"

ILLUSTRATIONS.—*First step*, James gave 7 cents for apples and 8 cents for peaches; how many cents did he spend? *Second step*, as many as the sum of 7 cents and 8 cents. *Third step*, 7 cents and 8 cents are 15 cents. *Fourth step*, hence, if James gave 7 cents for apples, and 8 cents for peaches, he spent 15 cents.

Again: *First step*, 4-fifths of 25 are how many times 6? *Second step*, as many times 6 as 6 is contained times in 4-fifths of 25. *Third step*, 1-fifth of 25 is 5, 4-fifths are 4 times 5, which are 20; 6 in 20 is contained 3 and 2-sixths times. *Fourth step*, therefore, 4-fifths of 25 are 3 and 2-sixths times 6?

As a means of keeping the attention of the class directed to each question, it will be proper for the instructor occasionally to read an example aloud, and, having allowed sufficient time for the answer, to call upon some one for the result, and then for the analysis. By this means, each one is obliged to solve the example, not knowing but that he may be required to answer it.

INTELLECTUAL ARITHMETIC.

SECTION I.—NUMERATION.

Pupils who have studied Ray's Arithmetic, First Book, may commence with Lesson IV, on page 14.

NUMERATION is naming Numbers.

Learn the name and form of these figures:

1   2   3   4   5   6   7   8   9   10.
One, two, three, four, five, six, seven, eight, nine, ten.

An Illustration of the Increase of Numbers.

●	·	·	·	·	·	·	·	·	1
●	●	·	·	·	·	·	·	·	2
●	●	●	·	·	·	·	·	·	3
●	●	●	●	·	·	·	·	·	4
●	●	●	●	●	·	·	·	·	5
●	●	●	●	●	●	·	·	·	6
●	●	●	●	●	●	●	·	·	7
●	●	●	●	●	●	●	●	·	8
●	●	●	●	●	●	●	●	●	9
●	●	●	●	●	●	●	●	●	10

SECTION II.—ADDITION.

LESSON I.

1. James had one apple, and his brother gave him one more: how many had he? Why?

Ans. Because 1 and 1 are 2.

2. Henry had two cents, and his sister gave him one more: how many had he? Why?

Ans. Because 2 and 1 are 3.

3. A boy had one marble, and found three more: how many did he then have? Why?

Ans. Because 1 and 3 are 4.

4. Thomas had 4 cents, and his mother gave him one more: how many cents had Thomas? Why?

5. Samuel had two cakes, and his father gave him two more: how many did he then have? Why?

6. Three oranges and two oranges are how many oranges? Why?

7. James had three apples, and his brother gave him three more: how many apples had James then? Why?

8. John had four plums, and his sister gave him two more: how many did he then have? Why?

9. Daniel had three cents; his brother gave him two, and his sister one: how many did he then have? Why?

10. Mary had four pears, and her brother gave her three more: how many did she then have? Why?

11. George recites daily four lessons perfectly, one imperfectly, and is absent from one recitation: how many daily lessons has he? Why?

12. How many fingers have you on one hand? How many on both? Why?

13. Ida had four cents; her mother gave her three more at one time, and one at another: how many did she then have? Why?

ADDITION.

14. Three cakes, and one cake, and two cakes, are how many cakes? Why?

15. Four cents, and three cents, and two cents, are how many cents? Why?

16. Five oranges, and one orange, are how many oranges? Why?

17. Henry had five cents, and his mother gave him two more: how many did he then have? Why?

18. Five boys, and one boy, and two boys, are how many boys? Why?

19. Oliver has five dollars, Henry three, and Samuel one: how many dollars have all together? Why?

20. Three peaches, and six peaches, are how many peaches? Why?

21. A lady paid one dollar for gloves, three dollars for a shawl, and five dollars for a dress: how much did she spend? Why?

22. Four cents, and three cents, and two cents, and one cent, are how many cents? Why?

23. If a man buy six pounds of sugar at one time, and four at another, how much does he buy? Why?

24. Seven oranges, and one orange, and two oranges, are how many? Why

25. George has two cents, his sister three cents, and his brother five cents: if they were all given to George how many would he have? Why?

26. How many are 4 and 3 and 3?

27. James has 4 cents, and Joseph 2 more than James: how many has Joseph? How many have both? Why?

LESSON II.

1. One and 1 are how many? 1 and 2? 3 and 1? 4 and 1? 1 and 3? 1 and 5? 1 and 6? 6 and 1? 1 and 7? 8 and 1? 9 and 1? 1 and 8?

2. Two and 2 are how many? 2 and 3? 2 and 2 and 1? 1 and 2 and 3? 4 and 1 and 2?

3. Four and 2 are how many? 4 and 3? 2 and 4 and 1? 4 and 1 and 2? 3 and 4? 5 and 4?

4. Six and 2 are how many? 6 and 1 and 2? 6 and 3? 6 and 4? 6 and 1 and 3? 1 and 2 and 6? 1 and 3 and 6?

5. Eight and 2 are how many? 8 and 3? 8 and 2 and 1? 2 and 8 and 1? 1 and 4 and 3? 8 and 4? 7 and 4? 6 and 7? 7 and 1 and 3? 1 and 6 and 2? 1 and 8 and 2 and 2?

6. Nine and 2 are how many? 9 and 3? 9 and 1 and 2? 8 and 1 and 3? 5 and 6? 1 and 4 and 5? 2 and 4 and 6? 3 and 4 and 5? 4 and 6 and 2? 5 and 4 and 2 and 1?

7. How many are 1 and 9? 7 and 3? 4 and 6? 9 and 1? 5 and 5? 6 and 4? 2 and 8? 3 and 7? 0 and 10? 8 and 2? 10 and 0? What two numbers added together make 10?

8. Which is the greater, 7 and 2, or 6 and 4? Why? one and 3 and 6, or 8 and 0 and 2? Why?

9. Begin at 4, and add 2 each time, up to 10.

10. Begin at 1, and add 3 each time, up to 13.

11. What two numbers added together will make 12? What three numbers?

12. Seven and 5 and 2 are how many?

13. One and 7 and 3 and 4 are how many?

14. If 3 be added to 3, and that sum to 4, what will be the result?

15. If you add 3, to 3 and 1 more, and then add 7, what will be the amount?

16. I have in one basket 8 dozen eggs, in another 4 dozen, in another 3 dozen: how many in all?

17. A lady bought 2 yards of tape for 3 cents, some pins for 1 cent, and received 2 cents change: how many cents had she at first?

18. Two and 1 more, and 3 and 4 more, are together how many?

19. One and 3 and 4 and 5 are how many? 5 and one and 3 and 4?

20. Two and 1 and 3, taken from a certain number, leave 2: what is that number?

21. A boy bought 3 cents worth of marbles, and 2 cents worth of candy, and received 5 cents change: how much money had he?

22. I bought 3 hams for 8 dollars, and 10 bushels of apples for 3 dollars: how much did I spend?

23. If 6 yards of cloth will make 2 coats, and 4 yards will make 2 pairs of pants, how many yards must I buy for 2 coats and 2 pairs of pants?

24. Oliver has 4 cents in one hand, 3 in the other, and 4 in his pocket: how many cents has he?

LESSON III.—ADDITION TABLE.

2 and	0 are	2	3 and	0 are	3	4 and	0 are	4
2 and	1 are	3	3 and	1 are	4	4 and	1 are	5
2 and	2 are	4	3 and	2 are	5	4 and	2 are	6
2 and	3 are	5	3 and	3 are	6	4 and	3 are	7
2 and	4 are	6	3 and	4 are	7	4 and	4 are	8
2 and	5 are	7	3 and	5 are	8	4 and	5 are	9
2 and	6 are	8	3 and	6 are	9	4 and	6 are	10
2 and	7 are	9	3 and	7 are	10	4 and	7 are	11
2 and	8 are	10	3 and	8 are	11	4 and	8 are	12
2 and	9 are	11	3 and	9 are	12	4 and	9 are	13
2 and	10 are	12	3 and	10 are	13	4 and	10 are	14
2 and	11 are	13	3 and	11 are	14	4 and	11 are	15
2 and	12 are	14	3 and	12 are	15	4 and	12 are	16

5	and	0	are	5	6	and	0	are	6	7	and	0	are	7
5	and	1	are	6	6	and	1	are	7	7	and	1	are	8
5	and	2	are	7	6	and	2	are	8	7	and	2	are	9
5	and	3	are	8	6	and	3	are	9	7	and	3	are	10
5	and	4	are	9	6	and	4	are	10	7	and	4	are	11
5	and	5	are	10	6	and	5	are	11	7	and	5	are	12
5	and	6	are	11	6	and	6	are	12	7	and	6	are	13
5	and	7	are	12	6	and	7	are	13	7	and	7	are	14
5	and	8	are	13	6	and	8	are	14	7	and	8	are	15
5	and	9	are	14	6	and	9	are	15	7	and	9	are	16
5	and	10	are	15	6	and	10	are	16	7	and	10	are	17
5	and	11	are	16	6	and	11	are	17	7	and	11	are	18
5	and	12	are	17	6	and	12	are	18	7	and	12	are	19
8	and	0	are	8	9	and	0	are	9	10	and	0	are	10
8	and	1	are	9	9	and	1	are	10	10	and	1	are	11
8	and	2	are	10	9	and	2	are	11	10	and	2	are	12
8	and	3	are	11	9	and	3	are	12	10	and	3	are	13
8	and	4	are	12	9	and	4	are	13	10	and	4	are	14
8	and	5	are	13	9	and	5	are	14	10	and	5	are	15
8	and	6	are	14	9	and	6	are	15	10	and	6	are	16
8	and	7	are	15	9	and	7	are	16	10	and	7	are	17
8	and	8	are	16	9	and	8	are	17	10	and	8	are	18
8	and	9	are	17	9	and	9	are	18	10	and	9	are	19
8	and	10	are	18	9	and	10	are	19	10	and	10	are	20
8	and	11	are	19	9	and	11	are	20	10	and	11	are	21
8	and	12	are	20	9	and	12	are	21	10	and	12	are	22

LESSON IV.

Note.—In the following exercises let the numbers be added both ways: thus, 2 and 3 are 5, 3 and 2 are 5; 4 and 3 are 7, 3 and 4 are 7; etc. In adding such numbers as 16 and 12; say 16 and 10 are 26, and 2 are 28.

1. Three and 8 are how many? 3 and 10?

2. Four and 7 are how many? 4 and 9? 4 and eleven? 4 and 10? 4 and 12?

ADDITION.

3. Five and 7 are how many? 5 and 6? 5 and 9? five and 12? 5 and 10? 5 and 8? 5 and 11?

4. Seven and 6 are how many? 7 and 10? 7 and eight? 7 and 12? 7 and 9? 7 and 11?

5. Nine and 11 are how many? 9 and 9? 9 and 12? nine and 7? 9 and 10? 9 and 8? 9 and 11?

6. Ten and 6 are how many? 10 and 8? 10 and ten? 10 and 12? 10 and 11? 10 and 5? 10 and 7? ten and 9?

7. Eleven and 2 are how many? 11 and 4? 11 and six? 11 and 5? 11 and 7? 11 and 9? 11 and 11? eleven and 10?

8. Twelve and 5 are how many? 12 and 4? 12 and 6? 12 and 8? 12 and 10? 12 and 7? 12 and 9? 12 and 11? 12 and 12?

9. Thirteen and 4 are how many? 13 and 6? 13 and 5? 13 and 7? 13 and 9? 13 and 10? 13 and 8? 13 and 11? 13 and 12?

10. Fourteen and four are how many? 14 and 6? 14 and 8? 14 and 5? 14 and 7? 14 and 10? 14 and 9? 14 and 11? 14 and 12?

11. Fifteen and 5 are how many? 15 and 7? 15 and 9? 15 and 4? 15 and 8? 15 and 10? 15 and 12? 15 and 11?

12. Sixteen and 4 are how many? 16 and 6? 16 and 8? 16 and 5? 16 and 7? 16 and 9? 16 and 11? 16 and 10? 16 and 12?

13. Seventeen and 6 are how many? 17 and 4? 17 and 7? 17 and 5? 17 and 9? 17 and 8? 17 and 10? 17 and 12? 17 and 11?

14. Eighteen and 10 are how many? 18 and 4? 18 and 7? 18 and 5? 18 and 8? 18 and 6? 18 and 9? 18 and 11? 18 and 12?

15. Nineteen and 5 are how many? 19 and 3? 19 and 2? 19 and 7? 19 and 9? 19 and 8? 19 and 10? 19 and 6? 19 and 12? 19 and 11?

16. How many are 9 and 2? 19 and 2? 29 and 2? 49 and 2? 69 and 2? 39 and 2? 59 and 2? 79 and 2? 99 and 2?

17. How many are 9 and 3? 3 and 19? 29 and 3? 3 and 49? 59 and 3? 3 and 39? 69 and 3? 3 and 79? 3 and 89? 99 and 3?

18. How many are 9 and 7? 29 and 7? 7 and 49? 39 and 7? 7 and 59? 79 and 7? 7 and 69? 89 and 7? 7 and 99?

19. How many are 9 and 8? 29 and 8? 49 and 8? 39 and 8? 8 and 69? 59 and 8? 79 and 8?

20. How many are 9 and 9? 19 and 9? 9 and 29? 49 and 9? 69 and 9? 59 and 9? 79 and 9? 89 and 9? 9 and 99?

21. How many are 8 and 3? 28 and 3? 48 and 3? 68 and 3? 88 and 3? 98 and 3?

22. How many are 8 and 7? 28 and 7? 7 and 38? 48 and 7? 68 and 7? 58 and 7? 88 and 7?

23. How many are 7 and 7? 17 and 7? 27 and 7? 47 and 7? 57 and 7? 37 and 7? 67 and 7? 87 and 7? 77 and 7? 97 and 7?

24. How many are 7 and 10? 17 and 10? 27 and 10? 47 and 10? 37 and 10? 57 and 10?

25. How many are 6 and 5? 16 and 5? 15 and 6? 26 and 5? 25 and 6? 24 and 6? 26 and 4? 36 and 6? 48 and 6? 45 and 6? 57 and 6? 59 and 6? 66 and 6? 75 and 6? 86 and 6?

26. How many are 13 and 8? 17 and 3? 23 and 8? 24 and 8? 33 and 8? 3 and 37? 8 and 43? 47 and 3? 7 and 53? 58 and 3? 67 and 3? 3 and 87? 97 and 3? 88 and 3? 3 and 98?

27. How many are 14 and 9? 9 and 24? 25 and 9? 9 and 34? 36 and 9? 9 and 44? 9 and 47? 54 and 9? 9 and 56? 9 and 64? 74 and 9? 9 and 72? 84 and 9? 86 and 9? 94 and 9?

ADDITION. 17

28. How many are 10 and 6? 6 and 21? 10 and 26? 46 and 10? 10 and 35? 10 and 55? 56 and 10? 10 and 66? 10 and 69? 76 and 10? 10 and 86? 96 and 10?

29. How many are 11 and 6? 11 and 16? 11 and 27? 25 and 11? 11 and 23? 31 and 11? 11 and 35? 37 and 11? 11 and 59? 46 and 11? 11 and 48? 52 and 11? 11 and 63?

30. Fifty-six and 1 are how many? 56 and 3? 56 and 5? 2 and 56? 4 and 56? 7 and 56? 56 and 6? 8 and 56? 56 and 9? 10 and 56?

31. Ninety-eight and 1? 3 and 98? 5 and 98? 98 and 2? 4 and 98? 98 and 6? 7 and 98? 98 and 8? 10 and 98? 98 and 9?

32. How many are 1 and 2 and 4 and 5? 5 and 1 and four and 2? 2 and 1 and 5 and 4?

33. How many are 12 and 8? 9 and 2 and 9? 4 and 2 and 9 and 5? 4 and 2 and 5 and 5 and 4? 8 and 4 and 6 and 2? 8 and 9 and 6?

34. If you add 3 to 3, and 4 to 4, and 5 to 5, and add those sums together, what number will you have?

35. If to 10 you add 6, 7, and 9, how much will you then have?

LESSON V.

NOTE.—The numbers in the following examples should be added aloud; and if preferred, the "Four step method" can be applied to all the questions.

Take for example the second question; *Four and 5 and 7 are how many?* As many as the sum of 4 and 5 and 7; 4 and 5 are 9, 9 and 7 are 16; therefore, 4 and 5 and 7 are 16.

1. Three and 6 and 4 are how many?
2. Four and 5 and 7 are how many?
3. Five and 6 and 2 are how many?
4. Six and 4 and 5 are how many?

5. Seven and 3 and 5 and 2 are how many?
6. Eight and 2 and 3 and 4 are how many?
7. Nine and 2 and 4 and 3 are how many?
8. Two and 9 and 5 and 4 are how many?
9. Three and 9 and 5 and 4 are how many?

10. Four and 8 and 3 and 5 and 2 and 6 and 3 and one are how many?

11. Five and 7 and 2 and 3 and 4 and 6 and 5 and 2 are how many?

12. Two and 4 and 3 and 5 and 6 and 2 and 7 and 4 are how many?

13. Three and 2 and 4 and 5 and 4 and 6 and 3 and seven and 5 are how many?

14. Four and 3 and 5 and 7 and 6 and 8 and 2 and 4 are how many?

15. Four and 9 and 3 and 5 and 6 and 7 and 8 and 9 are how many?

16. Five and 8 and 5 and 8 and 5 and 8 and 5 and 8 are how many?

17. Six and 8 and 7 and 3 and 5 and 4 and 7 and 1 and nine are how many?

18. Seven and 9 and 5 and 4 and 6 and 3 and 8 and five and 9 are how many?

19. Eight and 7 and 6 and 5 and 4 and 9 and 3 and seven and 8 are how many?

20. Nine and 6 and 7 and 4 and 5 and 3 and 8 and two and 9 are how many?

21. Seven and 6 and 5 and 8 and 7 and 9 and 8 and four and 9 and 8 are how many?

22. Nine and 8 and 7 and 5 and 8 and 9 and 5 and four and 7 and 3 and 9 and 8 are how many?

23. Twelve and 11 and 7 and 4 and 9 are how many?

24. Thirteen and 10 and 8 and 6 and 4 and 10 are how many?

ADDITION.

25. Fourteen and 16 and 7 and 5 and 9 and 8 and 9 and six and 4 are how many?

26. James gave 7 cents for apples, and 8 cents for peaches: how many cents did he spend?

27. Seven dollars, and 5 dollars, and 3 dollars, are how many dollars?

28. David had 11 books; he bought 7 more, and his brother gave him 5; how many had he then?

29. A man gave 13 dollars for a cart, 6 for a plow, and 5 for a harrow: how many dollars did he spend?

30. James has 8 marbles in one pocket, 5 in another, 6 in another, and 7 in another: how many in all?

31. If a pound of butter cost 18 cents, of lard 7 cents, and of cheese 9 cents, how much will all cost?

32. A man owes to one person 8 dollars, to another 5 dollars, to another 3 dollars, and to another 7 dollars: how much does he owe?

33. A boy gave 19 cents for a spelling-book, 8 cents for a slate, and 6 cents for pencils: how many cents did he spend?

34. A drover bought hogs as follows: of one man 17, of another 9, of another 7, of another 8: how many did he buy?

35. A little girl gave 10 cents for thread, 7 cents for pins, 6 cents for needles, and 9 cents for tape: how many cents did she spend?

36. William has 7 cents, Thomas 10 cents, David 9 cents, and Moses 8 cents: if the other boys give their cents to Moses, how many will he have?

37. The age of Thomas is 8 years; Frank, 5 years; and William is as old as both together: what is the sum of all their ages?

38. Joseph has 4 marbles, William has 2, and David has twice as many as Joseph: how many do they all have?

39. Begin with 2, and count one hundred by adding 2 successively. Thus, 2 and 2 are 4, and 2 are 6, and 2 are 8, and 2 are 10, and so on.

40. Begin with 3, and count ninety-nine by adding 3 successively. Thus, 3 and 3 are 6, and 3 are 9, and 3 are 12, and so on.

41. Begin with 4, and count one hundred by adding four successively.

42. Begin with 5, and count one hundred by adding five successively.

43. Begin with 6, and count one hundred and two by adding 6 successively.

44. Begin with 7, and count ninety-eight by adding seven successively.

45. Begin with 8, and count one hundred and four by adding 4 successively.

46. Begin with 9, and count ninety-nine by adding nine successively.

47. Begin with 1, and count one hundred by adding three successively.

48. Begin with 3, and count one hundred and three by adding 4 successively.

49. Begin with 2, and count one hundred and two by adding 5 successively.

50. Begin with 5, and count one hundred and seven by adding 6 successively.

51. Begin with 6, and count one hundred and four by adding 7 successively.

52. Begin with 7, and count one hundred and three by adding 8 successively.

53. Begin with 8, and count one hundred and seven by adding 9 successively.

54. If an apple cost 3 cents, and an orange cost 2 cents more than an apple, what will be the cost of 2 apples and one orange?

ADDITION.

55. James has 5 marbles; Henry, 2 more than James; and Samuel, as many as both James and Henry: how many have all?

56. Mary had a certain number of peaches; she gave 5 to her sister, 7 to her brother, and then had 10 remaining: how many had she?

57. A boy bought a sled for 20 cents, and paid 5 cents for repairing it: what was it then worth?

58. If a hat cost 4 dollars, and a coat as much as three hats, what do they both cost?

59. How many times must 10 be added to fifteen, to make fifty-five?

60. A boy's father gave him 5 cents; his mother, one cent more than his father; and his brother, two cents more than his father: how many did he then have?

61. What is the exercise of putting numbers together called? *Ans. Addition.*

62. What is simple Addition?

Ans. Simple Addition is finding the SUM *of two or more numbers of the same denomination.*

63. What is meant by *same denomination?*

Ans. Same denomination means all of the same name; that is, all pounds, or all dollars, &c.

SECTION III.—SUBTRACTION.

LESSON I.

1. James had 2 apples, and gave 1 to his brother: how many had he left? *Ans.* 1. *Why? Because* 1 *and* 1 *are* 2.

2. Then 1 from 2 leaves how many?

3. Joseph had 3 apples and lost 1: how many had he left? *Ans.* 2. *Why? Because* 1 *and* 2 *are* 3.

4. Then 1 from 3 leaves how many?

5. Thomas had 4 cents, and gave 1 to Frank: how many had he left? Why?

6. Then 1 from 4 leaves how many?

7. One from 5 leaves how many? From 6? 7? 8? 9? 10?

8. John had 4 cents and gave his sister 2: how many had he left? Why?

9. Then 2 from 4 leaves how many?

10. James had 5 apples, and gave his brother 2: how many had he left? Why?

11. Then 2 from 5 leaves how many?

12. Two from 6 leaves how many? From 7? 8? 9? 10? 11?

13. Thomas had 5 cents and lost 3: how many had he left? *Ans.* 2. Why?

14. Then 3 from 5 leaves how many?

15. Three from 6 leaves how many? From 7? 8? 9? 10? 11? 12?

16. Joseph had 9 marbles and lost 4: how many had he left? Why?

17. Then 4 from 9 leaves how many?

18. Four from 10 leaves how many? From 11? 12? 13? 14? 15?

19. William had 10 apples and gave Joseph 5: how many had he left? Why?

20. Then 5 from 10 leaves how many?

21. Five from 11 leaves how many? From 12? 13? 14? 15? 16?

22. James had 11 marbles and lost 6: how many had he left? Why?

23. Then 6 from 11 leaves how many?

SUBTRACTION.

24. Six from 12 leaves how many? From 13? 14? 15? 16? 17? 18?

25. William had 12 cents and lost 7: how many had he left? Why?

26. Then 7 from 12 leaves how many?

27. Seven from 13 leaves how many? From 14? 15? 16? 17? 18? 19?

28. James had 13 apples and gave his sister 8: how many had he left? Why?

29. Then 8 from 13 leaves how many?

30. Eight from 14 leaves how many? From 15? 16? 17? 18? 19? 20?

31. Thomas had 13 apples and gave his sister 9: how many had he left? Why?

32. Then 9 from 13 leaves how many?

33. Nine from 14 leaves how many? From 15? 16? 17? 18? 19? 20?

34. Henry had 14 cents and lost 5: how many had he remaining?

35. Mary is 15 years old, and Anna is 8: how much older is Mary than Anna?

36. Sold a load of corn for 17 dollars; received for it a barrel of flour worth 6 dollars, and the rest in money: how much money did I receive?

37. A boy had 18 marbles and lost 10: how many had he then?

38. Nineteen is 11 more than what number?

39. Went shopping with 20 dollars; spent 10 dollars in one store, and 5 in another: how much had I left?

☞ In the present edition of this volume, a very few examples have been modified, the better to adapt them to a more perfect gradation. Such are indicated by the character, O. The omission of these examples in classes having books of the present and former edition, will obviate all confusion.

LESSON. II.—SUBTRACTION TABLE.

1 from	1	leaves	0	2 from	2	leaves	0	3 from	3	leaves	0
1 from	2	leaves	1	2 from	3	leaves	1	3 from	4	leaves	1
1 from	3	leaves	2	2 from	4	leaves	2	3 from	5	leaves	2
1 from	4	leaves	3	2 from	5	leaves	3	3 from	6	leaves	3
1 from	5	leaves	4	2 from	6	leaves	4	3 from	7	leaves	4
1 from	6	leaves	5	2 from	7	leaves	5	3 from	8	leaves	5
1 from	7	leaves	6	2 from	8	leaves	6	3 from	9	leaves	6
1 from	8	leaves	7	2 from	9	leaves	7	3 from	10	leaves	7
1 from	9	leaves	8	2 from	10	leaves	8	3 from	11	leaves	8
1 from	10	leaves	9	2 from	11	leaves	9	3 from	12	leaves	9
1 from	11	leaves	10	2 from	12	leaves	10	3 from	13	leaves	10
1 from	12	leaves	11	2 from	13	leaves	11	3 from	14	leaves	11
4 from	4	leaves	0	5 from	5	leaves	0	6 from	6	leaves	0
4 from	5	leaves	1	5 from	6	leaves	1	6 from	7	leaves	1
4 from	6	leaves	2	5 from	7	leaves	2	6 from	8	leaves	2
4 from	7	leaves	3	5 from	8	leaves	3	6 from	9	leaves	3
4 from	8	leaves	4	5 from	9	leaves	4	6 from	10	leaves	4
4 from	9	leaves	5	5 from	10	leaves	5	6 from	11	leaves	5
4 from	10	leaves	6	5 from	11	leaves	6	6 from	12	leaves	6
4 from	11	leaves	7	5 from	12	leaves	7	6 from	13	leaves	7
4 from	12	leaves	8	5 from	13	leaves	8	6 from	14	leaves	8
4 from	13	leaves	9	5 from	14	leaves	9	6 from	15	leaves	9
4 from	14	leaves	10	5 from	15	leaves	10	6 from	16	leaves	10
4 from	15	leaves	11	5 from	16	leaves	11	6 from	17	leaves	11
7 from	7	leaves	0	8 from	8	leaves	0	9 from	9	leaves	0
7 from	8	leaves	1	8 from	9	leaves	1	9 from	10	leaves	1
7 from	9	leaves	2	8 from	10	leaves	2	9 from	11	leaves	2
7 from	10	leaves	3	8 from	11	leaves	3	9 from	12	leaves	3
7 from	11	leaves	4	8 from	12	leaves	4	9 from	13	leaves	4
7 from	12	leaves	5	8 from	13	leaves	5	9 from	14	leaves	5
7 from	13	leaves	6	8 from	14	leaves	6	9 from	15	leaves	6
7 from	14	leaves	7	8 from	15	leaves	7	9 from	16	leaves	7
7 from	15	leaves	8	8 from	16	leaves	8	9 from	17	leaves	8
7 from	16	leaves	9	8 from	17	leaves	9	9 from	18	leaves	9
7 from	17	leaves	10	8 from	18	leaves	10	9 from	19	leaves	10
7 from	18	leaves	11	8 from	19	leaves	11	9 from	20	leaves	11

SUBTRACTION.

.To TEACHERS.—Instead of requiring pupils to recite the Subtraction Table regularly, some of the best instructors omit it, and connect the exercises with addition, thus:

1 and 2 are 3	2 and 3 are 5
2 and 1 are 3	3 and 2 are 5
1 from 3 leaves 2	2 from 5 leaves 3
2 from 3 leaves 1	3 from 5 leaves 2

In case the Table should not be omitted, the above exercises will be found very profitable.

1. Two from 7 leaves how many?

2 from 12?	3 from 8?	3 from 10?
4 from 9?	4 from 12?	4 from 13?
5 from 14?	5 from 17?	6 from 11?
6 from 15?	6 from 18?	7 from 11?
7 from 15?	7 from 17?	7 from 19?
8 from 13?	8 from 15?	8 from 17?
9 from 13?	9 from 15?	9 from 17?

2. What number must be added to 8 to make 10?

To 12 to make 15?	To 9 to make 15?
To 17 to make 20?	To 16 to make 20?
To 14 to make 19?	To 19 to make 21?
To 13 to make 16?	To 12 to make 20?
To 11 to make 20?	To 13 to make 20?

LESSON III.

1. A boy gave 9 cents for a spelling-book, worth only 7 cents: how much did he pay for it more than it was worth? Why?

2. A man having 16 dollars lost 12: how many had he left? Why?

3. Bought a book for 12 cents, and a top for 7 cents: how much did the book cost more than the top?

4. Thomas had 18 cents given him by two boys; one gave 9: how many did the other give?

5. Bought a book for 14 cents, and gave the shopkeeper 20 cents: how much change did he return me?

6. William has 19 apples; in one pocket he has 15: how many are in the other?

7. A man has 25 miles to travel: when he has gone nineteen, how many will he still have to travel?

8. A boy gave 24 cents for a book, and sold it for sixteen cents: how much did he lose?

9. James had 24 marbles; he gave 19 to his brother: how many had he left?

10. A man bought a horse for 19 dollars, and sold him for 27 dollars: how much did he gain?

11. A man owing 26 dollars, paid 18: how many did he then owe?

12. Frank had 26 cents given him by William and Thomas. William gave him 17: how many did Thomas give? How many more did William give than Thomas?

13. If you had 10 apples, and should give 2 to John, and 6 to your sister, how many would you have left?

14. Abel had 36 cents, and his mother gave him enough to make 40: how many did she give him?

15. George had 40 marbles; he lost 20, and found five: how many did he then have?

16. A man had 100 barrels of flour; he sold 50, and afterward bought 10: how many did he then have?

17. A farmer had 35 bushels of grain; a part having been wasted, he found there were but 22 bushels remaining: how much was wasted?

18. John's father is 36 years old; John is 12: in how many years will he be as old as his father now is?

19. I had 65 cents; spent 20 cents for a book and ten for a slate: how many had I left?

20. If you take 10 from the sum of two numbers there will be 8 left: what is their sum?

21. If you take 16 from the difference of two numbers there will remain 12: what is their difference?

22. The sum of two numbers is 20: what number must be added to make their sum 30?

23. The sum of two numbers is 16 more than their difference; if their difference is 4, what is their sum?

24. The greater of two numbers is 12, and their difference 5: what is the less?

25. The sum of two numbers is 21; the less number is eight: what is the greater?

26. If you take one number from another, what is the operation called? *Ans. Subtraction.*

27. If you add one number to another, what is the operation called? *Ans. Addition.*

28. In what respect do Addition and Subtraction differ? *Ans. One is exactly the reverse or opposite of the other.*

29. When you take one number from another, what do you call that which is left?
Ans. Difference or Remainder.

30. The remainder and less number, added together are always equal to what? *Ans. The greater number.*

SECTION IV.—REVIEW.

LESSON I.

1. James had 13 marbles; he gave 2 to Henry, and three to Thomas: how many had he left?

ANALYSIS.—*He had as many left as the difference between* 13 *marbles, and the sum of* 2 *marbles and* 3 *marbles; the sum of* 2 *marbles and* 3 *marbles is* 5 *marbles;* 5 *from* 13 *leaves* 8 *marbles; therefore, if James had* 13 *marbles, and gave* 2 *to Henry and* 3 *to Thomas, he had* 8 *left.*

2. A merchant had 40 barrels of flour; he sold to one man 9, to another 21: how many had he left?

3. On Christmas day, William had 36 cents given him; he spent 6 cents for apples, 9 cents for cakes, and 10 cents for candy: how many had he left?

4. A man paid 30 dollars for a horse, the keeping cost 9 dollars, and he sold him for 29 dollars: how many dollars did he lose?

5. A man having 34 dollars, bought a barrel of molasses for 15 dollars, and a bag of coffee for 10 dollars: how many dollars had he left?

6. A grocer bought some oranges for 9 dollars, some lemons for 7 dollars, some prunes for 5 dollars, and some figs for 9 dollars, and then sold them for 41 dollars: how much did he gain?

7. A lady bought a comb for 25 cents, some pins for 10 cents, tape for 7 cents, thread for 6 cents, and toy books for 5 cents; she gave 60 cents to the shopkeeper: how much change ought she to receive?

8. Two boys commenced playing marbles; each had 18 when they began; when they quit, one had 25: how many had the other?

9°. Thomas has 7 marbles, David 5, and Moses 11; how many have they altogether? How many have Moses and David together more than Thomas?

10°. Three boys commence playing marbles: Thomas had 20, David 10, and Moses 4; when they quit, David had 6 and Moses 8: how many had Thomas?

11°. A farmer had 24 sheep: 9 of them were killed by wolves, 5 of them were stolen, and 6 he sold: how many had he left?

12. A grocer bought sugar for 12 dollars, flour for six dollars, and coffee for 5 dollars; he sold the whole for 30 dollars: how much did he make?

13. A lady had 50 cents; she spent 25 cents for butter, and 10 cents for eggs: how much had she left?

REVIEW. 29

14. A man is indebted to A, 5 dollars; to B, 6 dollars; and to C, 10 dollars: he has cash to the amount of 20 dollars, and goods valued at 10 dollars: should he pay his debts, how much would he be worth?

LESSON II.

1. Four and 3, less 2, are how many?

ANALYSIS.—*As many as the difference between* 2, *and the sum of* 4 *and* 3: 4 *and* 3 *are* 7; 7 *less* 2 *are* 5; *therefore,* 4 *and* 3 *less* 2 *are* 5.

2. Five and 6 and 2, less 8, are how many?
3. Seven and 4 and 3, less 5, are how many?
4. Eight and 5 and 4, less 3, are how many?
5. Two and 3 and 5, less 7, are how many?
6. Six and 4 and 3, less 6, are how many?
7. Six and 3 and 5, less 7, are how many?
8. Seven and 4 and 6, less 5, are how many?
9. Eight and 5 and 4 and 6, less 5, are how many?
10. Four and 7 and 6 and 5, less 8, are how many?
11. Five and 8 and 4 and 9, less 7, are how many?
12. Seven and 5 and 8 and 5, less 6, are how many?
13. Eight and 4 and 3 and 9, less 5, are how many?

14. Nine and 5 and 8 and 3 and 1, less 7, are how many?

15. Eight and 6 and 5 and 2 and 4 and 3, less 4, are how many?

16. Seven and 4 and 8 and 5 and 6 and 2 and 5, less eight, are how many?

17. Nine and 7 and 5 and 3 and 6 and 8 and 7, less six, are how many?

18. Eleven and 4 and 6 and 5 and 7 and 9, less 3, are how many?

19. Twelve and 5 and 7 and 6 and 8 and 3 and 4, less 7, are how many?

20. Twelve and 7 and 6 and 4 and 5 and 8 and 2, less 7, are how many?

21. Eleven and 6 and 5 and 3 and 6 and 8 and 6, less 9, are how many?

22. Eleven and 7 and 5 and 4 and 5 and 9 and 6, less 8, are how many?

23. Eleven and 9 and 8 and 7 and 6 and 5 and 4, less 3, are how many?

24. Twelve and 9 and 7 and 6 and 5 and 3, less 7, are how many?

25. Thirteen and 4 and 5 and 7 and 8 and 6 and 9, less 5, are how many?

26. Thirteen and 7 and 3 and 9 and 8 and 6 and 5, less 8, are how many?

27. Eighteen and 9 and 10 and 8 and 7, less 9, are how many?

28. Twenty-one and 5 and 6 and 7 and 8 and 9 and 10 and 9, less 8, are how many?

29. Seventy, less 10 and 9 and 8 and 6 and 5 and 4 and 3 and 2 and 4 and 5 and 6, are how many?

LESSON III.

1. Henry had 24 cents, and spent all but 15: how many did he spend?

2. A man bought a cask of wine containing 27 gallons; after selling 10 gallons he found there were but 9 gallons remaining, the rest having leaked out: how much did he lose?

3. If from 20 you take 12 less 3, how many will remain?

4. If from the sum of 19 and 10, you take the difference between 17 and 10, what will be left?

5. A man owed 60 dollars: he paid at one time 20 dollars, and at another 30 dollars: he afterward borrowed 5 dollars: how much does he still owe?

6°. A man paid 38 dollars for a horse, and 20 for a colt: he afterward sold the colt for 10 dollars, and the horse for 65: how much did he make by the transaction?

7. Twenty-four less 8, and 12 less 5, are together how much less than 25?

8. Engaged to do a piece of work for 60 dollars: had an assistant 25 days at a dollar a day, and paid 20 dollars for materials: how much did I clear?

9. If from the sum of 8 and 9 and 10 and 11, you take the sum of 4 and 5 and 6 and 7, what will you have remaining?

10. A jeweler bought a watch for 40 dollars, a chain for 15 dollars, and a key for 3 dollars: he sold them for 63 dollars: what did he gain?

11. A drover bought sheep as follows: of one man 10; of another, 12; of another, 5; of another, 3: he sold at one time 15; and at another, 5: how many were left?

12. A gentleman having 40 dollars, purchased a suit of clothes: his pants cost 7 dollars; vest, 5 dollars; coat, 25 dollars: how much had he left?

13. What number must be added to 25, to make a sum 14 less than 45?

14. What number must be taken from 62, to give a result which shall be 12 more than 45?

15. If from the sum of 25 and 10 and 12, you take the difference between 28 and 19, what will remain?

16. A man bought a horse for 40 dollars: and after paying 15 dollars for keeping him, sold him for 75 dollars: how much did he make?

17. A gentleman engaged in trade with 75 dollars: after losing at one time 10 dollars, and at another 5, he gained 20 dollars: how much did he then have?

SECTION V.—MULTIPLICATION.

LESSON I.

1. A boy gave 2 cents for one lemon, and 2 cents for another: how many cents did he give for both?

2. How many, then, are 2 times 2? Why? *Because 2 and 2 are 4.*

3. A boy gave 3 cents for one peach and 3 cents for another: how many cents did he give for both?

4. How many, then, are 2 times 3? Why?

5. At 4 cents apiece, what will 2 pears cost?

6. How many are 2 times 4? 4 times 2? Why?

7. At 3 cents apiece, what will 3 peaches cost?

8. How many are 3 times 3? Why?

9. At 3 cents apiece, what will 4 apples cost?

10. How many are 4 times 3? Why?

11. At 3 cents apiece, what will 5 pears cost? How many are 5 times 3? 3 times 5?

12. At 4 cents apiece, what will 4 lemons cost? How many are 4 times 4?

13. At 5 dollars a yard, what will 4 yards of cloth cost? How many are 4 times 5? 5 times 4?

14. At 6 dollars a barrel, what will 4 barrels of flour cost? How many are 4 times 6? 6 times 4?

15. At 5 cents apiece, what will 5 oranges cost? How many are 5 times 5?

16. At 6 cents a yard, what will 5 yards of tape cost? How many are 5 times 6? 6 times 5?

17. At 6 cents apiece, what will 6 oranges cost? How many are 6 times 6?

18. At 7 cents a yard, what will 2 yards of tape cost? How many are 2 times 7? 7 times 2?

MULTIPLICATION. 33

19. At 7 cents apiece, what will 3 lemons cost? How many are 3 times 7? 7 times 3?

20. At 7 cents apiece, what will 4 oranges cost? How many are 4 times 7? 7 times 4?

21. If 1 marble is worth 7 apples, how many apples are worth 5 marbles? How many are 5 times 7?

22. If 1 peach is worth 8 apples, how many apples are worth 2 peaches? How many are 2 times 8?

23. If 1 orange cost 8 cents, how many cents will three oranges cost? How many are 3 times 8?

24. If 1 barrel of flour cost 8 dollars, how many dollars will 4 barrels cost? How many are 4 times 8?

25. If 1 orange is worth 8 apples, how many apples are 5 oranges worth? How many are 5 times 8?

26. At 9 cents apiece, what will 2 oranges cost? How many are 2 times 9? 9 times 2?

27. At 9 cents a yard, what will 3 yards of ribbon cost? How many are 3 times 9? 9 times 3?

28. At 10 cents a quart, what will 3 quarts of chestnuts cost? How many are 3 times 10?

29. If 1 yard of cloth cost 11 dollars, what will 3 yards cost? How many are 3 times 11?

30. At 10 cents a bunch, what will 4 bunches of grapes cost? How many are 4 times 10?

31. How many are 5 times 10? 10 times 5?

SUGGESTION.—The exercises given under the subtraction table can be extended to multiplication. One great use of such exercises is to awaken an interest in the recitation.

Take, for example, the numbers 5 and 6. Five and 6 are 11; 6 and 5 are 11; 5 from 11 leaves 6; 6 from 11 leaves 5; 5 times 6 are 30; 6 times 5 are 30.

LESSON II.—MULTIPLICATION TABLE.

1 time 1 is 1			1 time 2 is 2			1 time 3 is 3		
2 times 1 are 2			2 times 2 are 4			2 times 3 are 6		
3 times 1 are 3			3 times 2 are 6			3 times 3 are 9		
4 times 1 are 4			4 times 2 are 8			4 times 3 are 12		
5 times 1 are 5			5 times 2 are 10			5 times 3 are 15		
6 times 1 are 6			6 times 2 are 12			6 times 3 are 18		
7 times 1 are 7			7 times 2 are 14			7 times 3 are 21		
8 times 1 are 8			8 times 2 are 16			8 times 3 are 24		
9 times 1 are 9			9 times 2 are 18			9 times 3 are 27		
10 times 1 are 10			10 times 2 are 20			10 times 3 are 30		
11 times 1 are 11			11 times 2 are 22			11 times 3 are 33		
12 times 1 are 12			12 times 2 are 24			12 times 3 are 36		
1 time 4 is 4			1 time 5 is 5			1 time 6 is 6		
2 times 4 are 8			2 times 5 are 10			2 times 6 are 12		
3 times 4 are 12			3 times 5 are 15			3 times 6 are 18		
4 times 4 are 16			4 times 5 are 20			4 times 6 are 24		
5 times 4 are 20			5 times 5 are 25			5 times 6 are 30		
6 times 4 are 24			6 times 5 are 30			6 times 6 are 36		
7 times 4 are 28			7 times 5 are 35			7 times 6 are 42		
8 times 4 are 32			8 times 5 are 40			8 times 6 are 48		
9 times 4 are 36			9 times 5 are 45			9 times 6 are 54		
10 times 4 are 40			10 times 5 are 50			10 times 6 are 60		
11 times 4 are 44			11 times 5 are 55			11 times 6 are 66		
12 times 4 are 48			12 times 5 are 60			12 times 6 are 72		
1 time 7 is 7			1 time 8 is 8			1 time 9 is 9		
2 times 7 are 14			2 times 8 are 16			2 times 9 are 18		
3 times 7 are 21			3 times 8 are 24			3 times 9 are 27		
4 times 7 are 28			4 times 8 are 32			4 times 9 are 36		
5 times 7 are 35			5 times 8 are 40			5 times 9 are 45		
6 times 7 are 42			6 times 8 are 48			6 times 9 are 54		
7 times 7 are 49			7 times 8 are 56			7 times 9 are 63		
8 times 7 are 56			8 times 8 are 64			8 times 9 are 72		
9 times 7 are 63			9 times 8 are 72			9 times 9 are 81		
10 times 7 are 70			10 times 8 are 80			10 times 9 are 90		
11 times 7 are 77			11 times 8 are 88			11 times 9 are 99		
12 times 7 are 84			12 times 8 are 96			12 times 9 are 108		

MULTIPLICATION. 35

1 time 10 is 10	1 time 11 is 11	1 time 12 is 12
2 times 10 are 20	2 times 11 are 22	2 times 12 are 24
3 times 10 are 30	3 times 11 are 33	3 times 12 are 36
4 times 10 are 40	4 times 11 are 44	4 times 12 are 48
5 times 10 are 50	5 times 11 are 55	5 times 12 are 60
6 times 10 are 60	6 times 11 are 66	6 times 12 are 72
7 times 10 are 70	7 times 11 are 77	7 times 12 are 84
8 times 10 are 80	8 times 11 are 88	8 times 12 are 96
9 times 10 are 90	9 times 11 are 99	9 times 12 are 108
10 times 10 are 100	10 times 11 are 110	10 times 12 are 120
11 times 10 are 110	11 times 11 are 121	11 times 12 are 132
12 times 10 are 120	12 times 11 are 132	12 times 12 are 144

EXERCISES ON THE TABLE.

How many are 4 times 7? 7 times 4? 9 times 3? 6 times 5? 8 times 2? 7 times 5? 6 times 7? 7 times 8?

Eight times 6? 6 times 9? 9 times 7? 8 times 5? 7 times 3? 9 times 8? 7 times 7?

Nine times 6? 8 times 7? 7 times 6? 9 times 9? 7 times 9? 10 times 5? 4 times 11? 12 times 3? 8 times 8? 6 times 11? 7 times 10? 8 times 9?

How many are 2 times 2? 3 times 2? 4 times 2? 3 times 3? 5 times 2? 4 times 3? 7 times 2? 5 times 3? 4 times 4? 9 times 2?

Five times 4? 3 times 7? 11 times 2? 8 times 3? 5 times 5? 9 times 3? 7 times 4? 10 times 3? 8 times 4? 11 times 3?

Seven times 5? 9 times 4? 4 times 10? 7 times 6? 11 times 4? 9 times 5? 8 times 6?

Seven times 7? 10 times 5? 9 times 6? 11 times 5? 8 times 7? 5 times 12? 9 times 7? 8 times 8? 11 times 6?

Ten times 7? 9 times 8? 11 times 7? 8 times 10? 9 times 9? 7 times 12? 8 times 11? 9 times 10? 8 times 12? 11 times 9? 10 times 10?

LESSON III.

1. At 2 cents each, what will 7 oranges cost?

ANALYSIS.—*Seven oranges will cost 7 times as much as 1 orange. If 1 orange cost 2 cents, 7 oranges will cost 7 times 2 cents, which are 14 cents; therefore, 7 oranges at 2 cents each will cost 14 cents.*

2. At 7 cents each, what will 3 melons cost?

3. At 6 cents a dozen, what cost 5 dozen apples?

4. At 6 cents a pound, what cost 7 pounds of beef?

NOTE.—The dollar sign, $, will now be used in place of the word dollar: thus, $5, $6; read 5 dollars, 6 dollars.

5. At $6 a pound, what cost 8 pounds of opium?

6. At $3 a barrel, what cost 9 barrels of cider?

7. At $4 a pair, what cost 7 pairs of boots?

8. At 8 cents a dozen, what cost 10 dozen pens?

9. What cost 6 yards of cloth at $7 a yard?

10. What cost 8 barrels of flour at $5 a barrel?

11. If a man travel 7 miles an hour, how far will he travel in 8 hours?

12. On a chessboard are 8 rows of squares, and 8 squares in each row: how many squares on the board?

13. An orchard has 11 rows of trees, and 7 trees in each row: how many trees in the orchard?

14. In 1 cent are 10 mills; how many mills are there in 3 cents? In 4? In 5? 6? 7? 8?

15. In 1 pint are 4 gills; how many gills are there in 2 pints? In 3? In 4? 5? 6? 7? 8?

16. In 1 bushel are 4 pecks; how many pecks are there in 2 bushels? In 3? In 4? 5? 6? 7?

17. In 1 peck are 8 quarts; how many quarts are there in 2 pecks? In 3? In 4? 5? 6? 7?

MULTIPLICATION.

18. In 1 bushel how many quarts? Why?
19. What will 9 yards of cloth cost at $6 a yard?
20. What will 9 oranges cost at 8 cents each?
21. Two men start from the same place and travel in opposite directions: one travels 2 miles an hour, the other 4 miles: how far will they be apart at the end of 1 hour? At the end of 2 hours? 3 hours?
22. If 2 men can do a job of work in 3 days, how many days will it take 1 man to do it?

ANALYSIS.—*It will require 1 man twice as long as 2 men. If it take 2 men 3 days, it will take 1 man twice 3 days, which are 6 days; therefore, if 2 men do a job of work in 3 days, 1 man will do it in 6 days.*

23. If 3 men can do a piece of work in 4 days, in how many days can 1 man do it?
24. If 4 men can do a piece of work in 6 days, in how many days can 1 man do it?
25. If a quantity of bread serve 8 men 4 days, how many days will it serve 1 man?
26. If a man can earn $6 in 1 week, how many dollars can he earn in 8 weeks?
27. A person has a job of work which 6 men can do in 9 days; but it is necessary to do it in one day: how many men must be employed?
28. If 2 barrels of cider last 6 persons 4 weeks, how many weeks will it last 1 person?
29. If $9 worth of provisions last 8 persons 11 days, how many persons will it last 1 day?
30. If the interest of $1 is 6 cents a year, what will be the interest for 2 years? For 3 years? For 4? 5? For 6? 7? 8? 9? 10?
31°. I bought 6 barrels of apples at $2 a barrel, and 4 barrels of sugar at $11 a barrel: how much did they both cost?

2d Bk. 3

32. Four and 4 are 8, and 4 are 12, and 4 are 16, and 4 are 20: here we find that 4 taken five times makes 20. What is this operation called? *Ans. Addition.*

33. When we say 5 times 4 make 20, what do we call the operation? *Ans. Multiplication.*

34. How, then, would you define Multiplication? *Ans. Multiplication is a short method of performing several additions of the same number.*

35. The number to be multiplied, 4, is called the *multiplicand;* the number you multiply by, 5, the *multiplier;* and the answer, 20, the *product.*

LESSON IV.

1. Bought 2 apples at 2 cents each, 2 pears at 3 cents each, and an orange for 5 cents: what did they cost?

2. Two men start from the same place and travel in the same direction; one, 5 miles an hour; the other, 7 miles: how far will they be apart in 10 hours?

3. If, in the above question, the men travel in opposite directions, how far will they be apart in 12 hours?

4. A lady went shopping with $15; she bought 4 yards of cloth at $2 a yard; 2 pairs of gloves at $1 a pair; and a shawl for $2: what did they cost, and how much had she left?

5. A man bought 4 peaches at 5 cents each, 3 pears at 3 cents each, and 2 pints of chestnuts at 5 cents a pint: how much did they cost?

6. What will be the sum of 3 and 9 and 7, less the sum of 8 and 6 and 1?

7. If a man earn 5 shillings a day, and a boy 3 shillings, how much will both earn in 7 days?

8. A drover gave $10 and 7 sheep, valued at $4 a head, for a cow and calf: how much did they cost?

MULTIPLICATION.

9°. A merchant sold cloth at $7 a yard: a tailor bought of this cloth, at one time, 5 yards, and at another, 3 yards: what was the amount of his bill?

10°. Two brothers, Henry and Rufus, each received for their work 3 dimes a day: how much did they both receive for 6 days' work?

11. If 12 horses can be sustained in a pasture 10 months, how many horses will it feed 1 month?

12. What is 3 times the difference between 15, and the sum of 5 and 2?

13. The sum of two numbers is 23; the smaller is 11: what is 5 times the larger?

14. The difference between two numbers is 7: if the larger be 12, what will 8 times the smaller be?

15. If a boy buy apples at 1 cent each, and sell them for 3 cents each, what would he make if he purchase 10 cents' worth of apples?

16. George bought a book for 50 cents and sold it for $1: what would he have made, had he bought 2 books, and sold them at the same rate as the first?

17°. Albert has 5 times two marbles less than 50, and Edward has 5 times two more than 50: how many has each?

18. If $1 gain $3 in a year, what will $12 gain in double the time?

19. A man bought a cask of wine containing 20 gallons, at $1 a gal.; 5 gal. having leaked out, he sold the remainder at $2 a gal.: how much did he make?

20. Two men start from the same place, at the same time, and travel the same way; if they travel at the same rate, how far will they be apart at the end of 10 hours?

If one goes 10 miles an hour, and the other 7, how far will they be apart in 7 hours?

If they go in opposite directions, each at the rate of 5 miles per hour, how far will they be apart in 9 hours?

21. A stage starts from a certain town, and travels at the rate of 8 miles per hour: at the same time, another starts from the same place, and travels in the same direction, 4 miles per hour: how far will they be apart at the end of 12 hours?

22. A grocer bought 10 pounds of tea at 7 shillings a pound; after using 3 pounds, he sold the remainder at 10 shillings a pound: how much did the 3 pounds which he used cost him, in the end?

23. Bought 6 bushels of corn at 5 dimes a bushel; sold 4 bushels at 6 dimes a bushel, and 2 bushels at 4 dimes a bushel: how much did I make?

24. If an orange cost 5 cents, and an apple 2 cents, what will 2 oranges and 4 apples cost?

25. If pork is 8 cents, and beef 10 cents a pound, what cost 7 pounds of pork and 6 pounds of beef?

26. If an orange cost 5 times as much as an apple, how much more will 6 oranges cost than 25 apples, if an apple is worth 1 cent?

27. If a pound of sugar cost 5 cents, and a pound of coffee 3 times as much, less 3 cents, what will be the cost of 3 pounds of sugar and 2 of coffee?

*28°. Bought, at one time, 5 yards of muslin at 10 cents a yard; at another, 10 yards at 5 cents a yard: how much did it all cost?

29. When salt is 4 cents a quart, and molasses 3 times as much lacking 2 cents, what would be the cost of 3 quarts of molasses and 2 of salt?

30. If a man earn $15 per week, and spend $11 a week, how much will he save in 3 weeks? How much can he save in two months of 4 weeks each?

---

* In the present edition of this volume, a very few examples have been modified, the better to adapt them to a more perfect gradation. Such are indicated by the character, o. The omission of these examples in classes having books of the present and former edition, will obviate all confusion.

SECTION VI.—DIVISION.

LESSON I.

1. At 1 cent each, how many cakes can you buy for 4 cents? 1 in 4 how many times? Why? *Ans. Because* 4 *times* 1 *are* 4.

2. At 2 cents each, how many apples can you buy for 4 cents?

ANALYSIS.—*You can buy as many apples as* 2 *cents are contained times in* 4 *cents;* 2 *cents are contained in* 4 *cents* 2 *times; therefore, at* 2 *cents each, you can buy* 2 *apples for* 4 *cents.*

Two in 4 how many times? Why? *Because* 2 *times* 2 *are* 4.

3. Among how many boys can 6 apples be divided, giving to each boy two apples? 2 in 6 how many times? Why? *Because* 3 *times* 2 *are* 6.

4. At 2 cents each, how many apples can you buy for 8 cents? 2 in 8 how many times? Why?

5. At 3 cents each, how many peaches can you buy for 6 cents? 3 in 6 how many times? Why?

6. At 3 cents each, how many pears can you buy for 9 cents? 3 in 9 how many times? Why?

7. At 2 cents each, how many cakes can you buy for 10 cents? 2 in 10 how many times? Why?

8. At 2 cents each, how many marbles can you buy for 14 cents? 2 in 14 how many times? Why?

9. At 5 cents each, how many lemons can you buy for 15 cents? 5 in 15 how many times? Why?

10. A boy has 16 marbles, and wishes to divide them into piles of 2 each: how many piles will there be? How many twos in 16? Why?

11. At 3 cents each, how many peaches can you buy for 18 cents? 3 in 18 how many times?

12. At 5 cents each, how many oranges can you buy for 20 cents? 5 in 20 how many times?

13. At $3 a yard, how many yards of cloth can you buy for $21? 3 in 21 how many times?

NOTE.—The dollar sign, $, is used in place of the word dollar: thus, $4, $7; read 4 dollars, 7 dollars.

14. A lady spent 22 cents for tape, at 2 cents a yard: how many yards did she buy?

15. At 6 cents each, how many oranges can you buy for 24 cents? How many at 8 cents each? 6 in 24 how many times? 8 in 24 how many times?

16. In an orchard of 25 apple trees there are 5 rows: how many trees in each row?

ANALYSIS.—*One tree in each row requires 5 trees; hence, there will be as many rows, as 5 trees are contained times in 25 trees; 5 trees in 25 trees, 5 times. Ans. 5 rows.*

17. If a man can travel 3 miles in an hour, how many hours will it take him to travel 27 miles? 3 in 27 how many times?

18. A man gave $28 for sheep, at $4 a head: how many did he buy? 4 in 28 how many times?

19. If you had 30 cents, how many marbles could you buy at 3 cents each? 3 in 30 how many times?

20. There are 32 cents on a table, in 4 piles: how many in each pile? 4 in 32 how many times?

21. In an orchard containing 35 apple trees, there are 5 rows: how many trees are there in each row? 5 in 35 how many times?

22. Six men receive $36 for a job of work: what is each man's share? 6 in 36 how many times?

DIVISION.

23. Four quarts make 1 gallon: how many gallons in 36 quarts? 4 in 36 how many times?

24. If a man travel 10 miles in 1 hour, in how many hours will he travel 40 miles?

25. Forty-two cents were divided equally among 6 boys: how many cents did each boy receive?

26. Forty-two are how many times 7?

27. If you divide 45 apples equally among 9 boys, how many apples will each boy receive?

LESSON II.—DIVISION TABLE.

2 in	2	1 time	3 in	3	1 time	4 in	4	1 time			
2 in	4	2 times	3 in	6	2 times	4 in	8	2 times			
2 in	6	3 times	3 in	9	3 times	4 in	12	3 times			
2 in	8	4 times	3 in	12	4 times	4 in	16	4 times			
2 in	10	5 times	3 in	15	5 times	4 in	20	5 times			
2 in	12	6 times	3 in	18	6 times	4 in	24	6 times			
2 in	14	7 times	3 in	21	7 times	4 in	28	7 times			
2 in	16	8 times	3 in	24	8 times	4 in	32	8 times			
2 in	18	9 times	3 in 27		9 times	4 in	36	9 times			
2 in	20	10 times	3 in	30	10 times	4 in	40	10 times			
2 in	22	11 times	3 in	33	11 times	4 in	44	11 times			
2 in	24	12 times	3 in	36	12 times	4 in	48	12 times			
5 in	5	1 time	6 in	6	1 time	7 in	7	1 time			
5 in	10	2 times	6 in	12	2 times	7 in	14	2 times			
5 in	15	3 times	6 in	18	3 times	7 in	21	3 times			
5 in	20	4 times	6 in	24	4 times	7 in	28	4 times			
5 in	25	5 times	6 in	30	5 times	7 in	35	5 times			
5 in	30	6 times	6 in	36	6 times	7 in	42	6 times			
5 in	35	7 times	6 in	42	7 times	7 in	49	7 times			
5 in	40	8 times	6 in	48	8 times	7 in	56	8 times			
5 in	45	9 times	6 in	54	9 times	7 in	63	9 times			
5 in	50	10 times	6 in	60	10 times	7 in	70	10 times			
5 in	55	11 times	6 in	66	11 times	7 in	77	11 times			
5 in	60	12 times	6 in	72	12 times	7 in	84	12 times			

8 in	8	1 time	9 in	9	1 time	10 in	10	1	time
8 in	16	2 times	9 in	18	2 times	10 in	20	2	times
8 in	24	3 times	9 in	27	3 times	10 in	30	3	times
8 in	32	4 times	9 in	36	4 times	10 in	40	4	times
8 in	40	5 times	9 in	45	5 times	10 in	50	5	times
8 in	48	6 times	9 in	54	6 times	10 in	60	6	times
8 in	56	7 times	9 in	63	7 times	10 in	70	7	times
8 in	64	8 times	9 in	72	8 times	10 in	80	8	times
8 in	72	9 times	9 in	81	9 times	10 in	90	9	times
8 in	80	10 times	9 in	90	10 times	10 in	100	10	times
8 in	88	11 times	9 in	99	11 times	10 in	110	11	times
8 in	96	12 times	9 in	108	12 times	10 in	120	12	times
11 in	11	1 time	11 in	55	5 times	11 in	99	9	times
11 in	22	2 times	11 in	66	6 times	11 in	110	10	times
11 in	33	3 times	11 in	77	7 times	11 in	121	11	times
11 in	44	4 times	11 in	88	8 times	11 in	132	12	times
12 in	12	1 time	12 in	60	5 times	12 in	108	9	times
12 in	24	2 times	12 in	72	6 times	12 in	120	10	times
12 in	36	3 times	12 in	84	7 times	12 in	132	11	times
12 in	48	4 times	12 in	96	8 times	12 in	144	12	times

NOTE.—The four simple operations of Arithmetic may now be combined in a single example: thus,

10 and 5 are 15
5 and 10 are 15
10 from 15 leaves 5
5 from 15 leaves 10

10 times 5 are 50
5 times 10 are 50
10 in 50 5 times
5 in 50 10 times

LESSON III.

1. Two in 12 how many times? 2 in 16? 2 in 24? 3 in 9? 3 in 15? 3 in 21? 3 in 27? 4 in 8? 4 in 20? 4 in 28? 4 in 36? 4 in 48?

2. Five in 15 how many times? 5 in 30? 5 in 45? 5 in 60? 6 in 18? 6 in 24? 6 in 36? 6 in 42? 6 in 54? 6 in 66?

DIVISION.

3. Seven in 14 how many times? 7 in 28? 7 in 42? 7 in 56? 7 in 63? 7 in 84? 8 in 24? 8 in 40? 8 in 56? 8 in 72? 8 in 96?

4. Nine in 18 how many times? 9 in 27? 9 in 45? 9 in 54? 9 in 63? 9 in 81? 9 in 108? 10 in 20? 10 in 60? 10 in 90? 10 in 100?

5. Eleven in 55 how many times? 11 in 77? 11 in 99? 11 in 110? 11 in 121? 12 in 24? 12 in 48? 12 in 60? 12 in 72? 12 in 96? 12 in 108? 12 in 120? 12 in 144?

6. If 12 peaches be divided equally among 3 children, how many will each have?

7. Four boys gave their sister 24 apples, each an equal number: how many did each give?

8. A mother divided 20 cents equally between her 2 little girls: how many did each receive?

9. Five books cost 35 cents: how much is that apiece?

10. A man has $40: if he spend $5 a week, how long will it last?

11. There are 3 feet in 1 yard: how many yards are there in 21 feet? In 27 feet? In 36 feet?

12. Four quarts make 1 gallon: how many gallons in 28 quarts? In 16? In 32? In 36? 44? 48?

13. If 5 apples are worth 1 pear, how many pears are worth 25 apples? 35 apples? 45 apples?

14. If 6 pears are worth an orange, how many oranges can you get for 30 pears? For 42 pears? For 54 pears? For 66 pears?

15. If 1 man do a piece of work in 42 days, how many days will it take 7 men?

16. If 1 man eat a certain quantity of provisions in 56 days, how many days will it last 7 men?

17. If 1 pipe empty a cistern in 63 hours, in how many hours will 9 pipes of the same size empty it?

18. Eight quarts make a peck: how many pecks in 24 quarts? In 40? In 56? In 72?

19. If hay is worth $9 a ton, how many tons can be bought for $27? For $45? For $54? For $63?

20. Ten men bought a horse for $60: how much did each one pay?

21. If 11 ounces cost 88 cents: what cost 1 ounce?

22. A man paid $108 for 12 Saxony sheep: how much was that apiece?

23. In an orchard there are 120 trees in 10 rows: how many trees in each row?

24. A man earns $144 in 12 weeks: how much is that a week? How much a day, allowing 6 working days to the week?

25. If 6 men earn $84 in 7 days, how much do they all earn in 1 day?

26. If 9 men earn $108 in 3 days, how much does 1 man earn? How much does each man earn in a day?

27. If 2 from 6 leaves 4, 2 from 4 leaves 2, and 2 from 2 leaves 0, how many times is 2 taken from 6? *Ans.* 3. What do you call the operation? *Subtraction.*

28. Two is contained in 6, 3 times; what do you call the operation? *Ans. Division.*

29. How would you define Division? *Ans. Division is subtracting the same number several times; or, finding how many times one number is contained in another.*

30. The number to be divided, 6, is called the *dividend;* the number you divide by, 2, the *divisor;* and the answer, 3, the *quotient.*

LESSON IV.

1. Twelve are how many times 2? 3? 4? 6?
2. Twenty-four are how many times 3? 6? 8? 12?
3. Seventy-two are how many times 12? 8? 6? 9?

DIVISION.

4. How many oranges at 5 cents each, must be given for 10 pears at 2 cents each?

5. A wheel is 10 feet in circumference; what distance will it move in making one revolution? how many revolutions will it make in going 120 feet?

6. An orchard contains 10 rows of trees, and 7 trees in a row; if there were but 5 rows, how many trees would there be in a row?

7. I have three times as many marbles as the sum of 1, 2, and 3, is contained times in 60: how many have I?

8. Bought 6 hats at $5 apiece, and 4 yards of cloth at $3 a yard; gave in exchange flour at $6 a barrel: how many barrels did it take?

9. If a man gain 6 miles in 3 hours, how long will it take to gain 24 miles?

10. Two times 6 are contained how many times in the sum of 36 and 12?

11. If 60 be divided by some number, the result will be 10: what is that number?

12. I have a number in my mind which, divided by 3, gives 2 times 6: what is the number?

13°. If I purchase lemons at the rate of 2 for 6 cents, and sell 7 for 28 cents, how much do I gain?

14. A man has a job of work which 9 men can perform in 2 days; he desires to complete it in 3 days: how many men must he employ?

15. Five times the sum of two numbers is equal to 60; if 7 is one of them, what is the other?

16. Henry has 6 dimes; Thomas twice as many less 2; and Samuel 3 times as many as Henry: how many have they together?

17°. If to the number of times 4 is contained in 12, you add 3, and subtract the result from 9, what will remain?

18. Five oranges were sold for 25 cents, and 10 cents were gained: what did each cost?

19. What number subtracted from 17, will leave double the remainder that 5 from 9 leaves?

20. A boy said that 10 taken from the number of apples he had, left twice as great a remainder as the difference between 1 dozen and 8: how many had he?

21. A certain number multiplied by 10, is 5 less than 45: what is that number?

22. If you multiply any number, 10, by any other number, 5, and divide the product by the same number, 5, what will be the result?

23. If 2 oranges are worth 5 apples, how many apples are worth 12 oranges?

ANALYSIS.—*As many times 5 apples, as 2 oranges are contained times in 12 oranges.*

24. One man goes 10 miles while another goes 7; when the first has gone 90 miles, how far will the second have gone?

25°. James earns 8 cents while John earns 12; when John has earned 60, how many has James earned?

26°. George learns 5 lessons while Charles learns 4: how many lessons will both have learned when Charles has learned 20?

27°. A man can earn $9 while a boy earns 5: how many dollars will both have earned when the man has earned $36?

28. How many times will the hammer of a common clock strike, from noon till 6 o'clock in the evening?

29. Two numbers added together make 30; if the greater number was 5 more, and the less 3 more, what would their sum then be?

30. William can count 11 while James counts 7: how many will James count while William is counting 77?

PARTS OF NUMBERS. 49

31. When flour was 5 cents a pound and sugar 11 cents, 10 pounds of flour and 5 pounds of sugar were given for a box of eggs at 5 cents a dozen: how many dozen were in the box?

32. The sum of two numbers is 24; if they were both equal to the greater, their sum would be 28: what are the numbers?

SEC. VII.—PARTS OF NUMBERS.

LESSON I.

If a *unit* or whole thing is divided into TWO equal parts, one of the parts is called ONE-HALF.

1. When an apple is divided into two equal parts, what is one part called? *Ans. One-half of an apple.*

2. How many halves in 1 apple? *Ans. Two halves.*

3. How many halves in 2 apples? In 3? In 4? In 5? In 6? In 7? In 8? In 9? In 10?

ANALYSIS.—*In* 2 *apples there are two times as many halves as there are in* 1 *apple; but in* 1 *apple there are* 2 *halves, and in* 2 *apples there will be* 2 *times* 2 *halves, or* 4 *halves; therefore, there are* 4 *halves in* 2 *apples.*

4. How many halves in 2 apples and one-half of an apple? In 3 apples and one-half? In 4 and one-half?

5. What do you mean by one-half of anything?

If any thing or any number is divided into THREE equal parts, one of the parts is called ONE-THIRD of the thing or number; two parts are called TWO-THIRDS; and three parts, THREE-THIRDS, or the whole.

6. When an apple, or any thing is divided into three equal parts, what is one part called? What two parts?

7. How many thirds in 1 apple? In 2? In 3? In 4? In 5? In 6? In 7? In 8? In 9?

8. How many thirds in 1 apple and one-third of an apple? In one apple and two-thirds of an apple?

9. How many thirds in two apples and one-third? In 2 apples and two-thirds?

10. What do you understand by one-third of any thing? By two-thirds? *Ans. That a whole thing has been divided into* 3 *equal parts and* 2 *of those parts taken.*

If any thing is divided into 4 equal parts, 1 of the parts is called ONE-FOURTH; two parts are called TWO-FOURTHS, and so on.

When any thing is divided into 5 equal parts, 1 of the parts is called ONE-FIFTH; two parts, TWO-FIFTHS; three parts, THREE-FIFTHS, and so on.

11. When an apple is divided into four equal parts, what is one part called? What two parts? *Ans. Two-fourths.* What three parts? *Ans. Three-fourths.*

12. How many fourths in 1 apple? In 2 apples? In 3 apples? In 4? In 5? In 6? In 7?

13. How many fourths in one apple and one-fourth? In 1 and two-fourths? In 1 and three-fourths?

14. In 2 apples and one-fourth how many fourths? In 2 apples and two-fourths? In 2 and three-fourths?

15. What are fourths often called? *Ans. Quarters.*

16. What do you understand by one-fourth of any thing? By two-fourths? By three-fourths?

17. When an apple is divided into 5 equal parts, what is one part called? What are 2 parts called? What 3 parts? What 4 parts? *Ans. Four-fifths.*

18. How many fifths in 1 apple? In 2? In 3? In 4? In 5? In 6? In 7? In 8? In 9?

19. How many fifths in 1 apple and 1-fifth of an apple? In 1 and 2-fifths? 1 and 3-fifths? 1 and 4-fifths?

20. How many fifths in 2 apples and 1-fifth? In 2 and 2-fifths? 2 and 3-fifths? 2 and 4-fifths?

PARTS OF NUMBERS.

21. What do you understand by 1-fifth of any thing? By 2-fifths? By 3-fifths?

22. When an apple is divided into 6 equal parts, what is 1 part called? What are 2 parts called? What 3 parts? What 4 parts? What 5 parts?

23. How many sixths in 1 apple? In 2 apples? In 3 apples? In 4? In 5? In 6? In 7? In 8?

24. How many sevenths in 1 apple? In 2 apples? In 3? In 4? In 5? In 6? In 7? In 8?

25. How many eighths in 1 apple? In 2? In 3? In 4? In 5? In 6? In 7? In 8? In 9?

26. How many ninths in 1 apple? In 2? In 3? In 4? In 5? In 6? In 7? In 8? In 9?

LESSON II.

1. If a yard of tape is worth 2 cents, what is one-half of it worth? *Ans. One cent.*

2. What is one-half of 2 cents? *Ans. One cent.* Why? *Ans. Because if you divide 2 cents into two equal parts, one of the parts is 1 cent.*

3. One is what part of 2? *Ans. One is the half of 2.*

4. If you can buy an apple for 2 cents, how many can you buy for 3 cents? *Ans. One apple and 1-half.*

5. Three are how many times 2? *Ans. Once 2 and 1-half of 2.*

6. Four are how many times 2?

7. If 2 cents will buy 1 yard of tape, how many yards will 5 cents buy?

8. Five are how many times 2? *Ans. Two times 2 and 1-half of 2.*

9. 6 are how many times 2? How many 2's?
10. 7 are how many times 2? How many 2's?
11. 8 are how many times 2? How many 2's?

12. 9 are how many times 2?   How many 2's?

13. 10 are how many times 2? How many 2's?

14. If an apple is worth 3 cents, what is 1-third of it worth? What are 2-thirds of it worth?

15. What is 1-third of 3? What are 2-thirds of 3?

16. If an orange is worth 3 cents, what part of it will 1 cent buy? What part will 2 cents buy?

17. One is what part of 3? *Ans. One is 1-third of 3.*

18. Two is what part of 3? *Ans. Two is 2-thirds of 3, or 2 times 1-third of 3.*

19. If a yard of cloth cost $3, how much can you buy for $4? How much for $5?

20. Four are how many times 3? *Ans. Once 3 and 1-third of 3.*

21. Five are how many times 3? *Ans. Once 3 and 2-thirds of 3.*

22. 6 are how many times 3?   7 are how many 3's?

23. 8 are how many times 3?   9 are how many 3's?

24. 10 are how many times 3? 11 are how many 3's?

25. If a lemon is worth 4 cents, what is 1-fourth of it worth?

26. What is 1-fourth of 4? What are 2-fourths of 4? What are 3-fourths of 4?

27. If you buy a yard of cloth for $4, what part of it can you buy for $1? For $2? For $3?

28. What part of 4 is 1?   *Ans. 1-fourth of 4.*

29. What part of 4 is 2?   *Ans. 2-fourths of 4.*

30. What part of 4 is 3?   *Ans. 3-fourths of 4.*

31. If a yard of tape cost 4 cents, how much can you buy for 5 cents? For 6 cents? For 7 cents?

32. Five are how many times 4? *Ans. Once 4 and one-fourth of 4.*

33. Six are how many times 4?   *Ans. Once 4 and two-fourths of 4.*

PARTS OF NUMBERS. 53

34. Seven are how many times 4? *Ans. Once 4 and three-fourths of 4.*

35. 8 are how many times 4?   9 are how many 4's?
36. 10 are how many times 4?   11 are how many 4's?
37. 12 are how many times 4?   13 are how many 4's?
38. 14 are how many times 4?   15 are how many 4's?
39. 16 are how many times 4?   17 are how many 4's?
40. 18 are how many times 4?   19 are how many 4's?
41. 20 are how many times 4?   21 are how many 4's?

42. If a melon is worth 5 cents, what is one-fifth of it worth? What are two-fifths worth? What are three-fifths worth? What are four-fifths worth?

43. What is 1-fifth of 5? 2-fifths of 5? 3-fifths of 5? 4-fifths of 5?

44. If a melon is worth 5 cents, what part of it can you buy for 1 cent? For 2 cents? For 3? For 4?

45. 1 is what part of 5?   *Ans. 1-fifth of 5.*
46. 2 is what part of 5?   *Ans. 2-fifths of 5.*
47. 3 is what part of 5?   *Ans. 3-fifths of 5.*
48. 4 is what part of 5?   *Ans. 4-fifths of 5.*

49. If flour is worth $5 a barrel, how many barrels can you buy for $6? For $7? For $8? For $9? For $10? For $12?

50. 6 are how many times 5? *Once 5 and 1-fifth of 5.*
51. 7 are how many times 5? *Once 5 and 2-fifths of 5.*
52. 8 are how many times 5? *Once 5 and 3-fifths of 5.*
53. 9 are how many times 5? 10 are how many times 5?
54. 11 are how many times 5? 12 are how many times 5?

55. If 1 barrel of apples cost $6, what is 1-sixth worth? What 2-sixths? What 3-sixths? 4-sixths? 5-sixths?

56. What is 1-sixth of 6? What are 2-sixths of 6? 3-sixths of 6? 4-sixths of 6? 5-sixths of 6?

57. One is what part of 6? 2 is what part of 6? 3 is what part of 6? 4 is what part of 6? 5?

58. How many yards of cloth at $6 a yard, can you buy for $7? For $8? For $9? For $10? For $11? For $12? For $13? For $14?

59. Seven are how many times 6? *Ans. One time 6 and 1-sixth of 6.*

60. 8 are how many times 6?   8 are how many 6's?
61. 9 are how many times 6?   10 are how many 6's?
62. 11 are how many times 6? 12 are how many 6's?
63. 13 are how many times 6? 14 are how many 6's?
64. 15 are how many times 6? 16 are how many 6's?

65. At 7 cents a yard for ribbon, what is 1-seventh worth? What are 2-sevenths worth? What are 3-sevenths? 4-sevenths? 5-sevenths? 6-sevenths?

66. What is 1-seventh of 7? What are 2-sevenths of 7? What are 3-sevenths? 4-sevenths? 5-sevenths? 6-sevenths?

67. One is what part of 7? 2 is what part of 7? 3 is what part of 7? 4 is what part of 7? 5? 6?

68. How many barrels of flour at $7 a barrel, can you buy for $8? For $9? For $10? For $11? For $12? For $13? For $14? For $15? For $16?

69. 8 are how many times 7?   9 are how many 7's?
70. 10 are how many times 7? 11 are how many 7's?
71. 12 are how many times 7? 13 are how many 7's?
72. 15 are how many times 7? 16 are how many 7's?
73. 17 are how many times 7? 18 are how many 7's?

74. If a melon cost 8 cents, what is 1-eighth worth? What are 2-eighths worth? 3-eighths worth? 4-eighths worth? 5-eighths? 6-eighths? 7-eighths?

75. What is 1-eighth of 8? What are 2-eighths of 8? 3-eighths of 8? 4-eighths? 5-eighths? 6-eighths? 7-eighths? 8-eighths? 9-eighths? 10-eighths?

PARTS OF NUMBERS.

76. One is what part of 8? 2 is what part of 8? 3 is what part of 8? 4 is what part of 8? 5 is what part of 8? 6 is what part of 8? 7?

77. How many gallons of beer, at 8 cents a gallon, can you buy for 9 cents? For 10 cents? For 11 cents? For 12 cents? For 13 cents? For 14 cents? For 15 cents? 16 cents? 17 cents? 18 cents?

78. 9 are how many times 8? *Once 8 and 1-eighth of 8.*
79. 10 are how many times 8? 11 are how many 8's?
80. 12 are how many times 8? 13 are how many 8's?
81. 14 are how many times 8? 15 are how many 8's?
82. 16 are how many times 8? 17 are how many 8's?

83. Eighteen are how many times 8? *Ans. Two times 8 and 2-eighths of 8.*

84. If an orange cost 9 cents, what is 1-ninth of it worth? What are 2-ninths worth? 3-ninths? 4-ninths? 5-ninths? 6-ninths? 7-ninths? 8-ninths?

85. What is 1-ninth of 9? What are 2-ninths of 9? What are 3-ninths of 9? 4-ninths of 9? 5-ninths of 9? 6-ninths of 9? 7-ninths of 9? 8-ninths of 9?

86. One is what part of 9? 2 is what part of 9? 3 is what part of 9? 4 is what part of 9? 5 is what part of 9? 6? 7? 8?

87. If cloth is $9 a yard, how much can you buy for $10? For $11? For $12? For $13? For $14? For $15? For $16? For $17? For $18? For $19? For $20?

88. If apples cost 10 cents a bushel, what is 1-tenth of a bushel worth? 2-tenths? 3-tenths? 4-tenths? 5-tenths? 6-tenths? 7-tenths? 8-tenths? 9-tenths?

89. One is what part of 10? 2 is what part of 10? 3 is what part of 10? 4 is what part of 10? 5 is what part of 10? 6 is what part of 10? 7 is what part of 10? 8 is what part of 10? 9? 11?

90. When cloth is $10 a yard, how much can you buy for $11? For $12? For $13? For $14? For $15? For $16? For $17? For $18? For $19?

91. What do you understand by 1-fifth of any thing? By 3-fifths? By 4-fifths?

92. What do you understand by 3-sevenths of any thing? By 2-eighths?

93. If I cut an apple into 6 equal parts, and give you 4, what part of the whole apple would I give you?

94. If I divide an orange into 8 equal parts, and give you 5, what part of the orange would I give you?

95. What is meant by 1-ninth of any thing? 2-ninths?

96. What is meant by 1-tenth of any thing? 2-tenths?

LESSON III.

1. If you had 7 cents, how many cakes could you buy at 2 cents each? *Ans. 3 cakes and 1-half.*

2. Seven are how many times 2? *Ans. 3 times 2 and 1-half of 2.*

3. If you had 11 cents, how many pears could you buy at 2 cents each? At 3 cents?

4. Eleven are how many times 2? How many times 3? *Ans. Three times 3 and 2-thirds of 3.*

5. If you had 15 cents, how many cakes could you buy at 2 cents each? At 3 cents? At 4 cents? At 5?

6. Fifteen are how many times 2? 3? 4? 5?

7. If you had 17 cents, how many oranges could you buy at 2 cents each? At 3 cents each? At 4 cents each? At 5 cents? At 6 cents? At 7 cents?

8. 24 are how many times 2? 3? 4? 5? 6? 7? 8?

9. 25 are how many times 2? 3? 4? 5? 6? 7? 8?

10. 26 are how many times 3? 4? 5? 6? 7? 8? 9?

11. 27 are how many times 3? 4? 5? 6? 7? 8? 9?

12. 28 are how many times 3? 4. 5? 6? 7? 8? 9?

PARTS OF NUMBERS. 57

13. 29 are how many times 3? 4? 5? 6? 7? 8? 9?
14. 30 are how many times 3? 4? 5? 6? 7? 8? 9?
15. 31 are how many times 3? 4? 5? 6? 7? 8? 9?
16. 32 are how many times 3? 4? 5? 6? 7? 8? 9?
17. 33 are how many times 3? 4? 5? 6? 7? 8? 9?
18. 34 are how many times 3? 4? 5? 6? 7? 8? 9?
19. 35 are how many times 3? 4? 5? 6? 7? 8? 9?
20. 36 are how many times 3? 4? 5? 6? 7? 8? 9?
21. 37 are how many times 3? 4? 5? 6? 7? 8? 9?
22. 38 are how many times 3? 4? 5? 6? 7? 8? 9?
23. 39 are how many times 4? 5? 6? 7? 8? 9?
24. 40 are how many times 4? 5? 6? 7? 8? 9?

25. If you had 41 cents, how many oranges could you buy at 4 cents each? At 5 cents each? At 6 cents? At 7 cents? At 8 cents? At 9 cents? At 10 cents?

26. 42 are how many times 4? 5? 6? 7? 8? 9?
27. 43 are how many times 4? 5? 6? 7? 8? 9?
28. 44 are how many times 4? 5? 6? 7? 8? 9?
29. 45 are how many times 4? 5? 6? 7? 8? 9?
30. 46 are how many times 4? 5? 6? 7? 8? 9?
31. 47 are how many times 4? 5? 6? 7? 8? 9?
32. 48 are how many times 4? 5? 6? 7? 8? 9?
33. 49 are how many times 4? 5? 6? 7? 8? 9?

34. If you had 56 cents, how many peaches could you buy at 5 cents each? At 6 cents each? At 7 cents? At 8 cents? At 9 cents? At 10 cents?

35. 56 are how many times 5? 6? 7? 8? 9? 10?
36. 57 are how many times 5? 6? 7? 8? 9? 10?
37. 58 are how many times 5? 6? 7? 8? 9? 10?
38. 59 are how many times 5? 6? 7? 8? 9? 10?
39. 60 are how many times 5? 6? 7? 8? 9? 10?

40. 61 are how many times 6? 7? 8? 9? 10?
41. 63 are how many times 6? 7? 8? 9? 10?
42. 65 are how many times 6? 7? 8? 9? 10?
43. 66 are how many times 6? 7? 8? 9? 10?
44. 68 are how many times 6? 7? 8? 9? 10?
45. 69 are how many times 6? 7? 8? 9? 10?

46. If 70 hours be required to perform a piece of work, in how many days can it be done by working 6 hours a day? By working 7 hours a day? 8 hours a day? 9 hours? 10 hours?

47. 71 are how many times 6? 7? 8? 9? 10?
48. 72 are how many times 6? 7? 8? 9? 10?
49. 74 are how many times 6? 7? 8? 9? 10?
50. 76 are how many times 6? 7? 8? 9? 10?
51. 77 are how many times 6? 7? 8? 9? 10?
52. 79 are how many times 7? 8? 9? 10? 11?
53. 80 are how many times 7? 8? 9? 10? 11?
54. 81 are how many times 7? 8? 9? 10? 11?
55. 82 are how many times 7? 8? 9? 10? 11?
56. 83 are how many times 7? 8? 9? 10? 11?
57. 85 are how many times 7? 8? 9? 10? 11?
58. 86 are how many times 7? 8? 9? 10? 11?
59. 87 are how many times 7? 8? 9? 10? 11?
60. 88 are how many times 7? 8? 9? 10? 11?
61. 89 are how many times 7? 8? 9? 10? 11?
62. 90 are how many times 7? 8? 9? 10? 11?
63. 91 are how many times 8? 9? 10? 11? 12?
64. 92 are how many times 8? 9? 10? 11? 12?
65. 93 are how many times 8? 9? 10? 11? 12?
66. 94 are how many times 8? 9? 10? 11? 12?
67. 95 are how many times 8? 9? 10? 11? 12?
68. 96 are how many times 8? 9? 10? 11? 12?

PARTS OF NUMBERS.

69. 98 are how many times 8? 9? 10? 11? 12?
70. 99 are how many times 8? 9? 10? 11? 12?
71. 100 are how many times 8? 9? 10? 11? 12?

SECTION VIII.
LESSON I.

1. At 2 cents each, what will 2 apples and 1-half of an apple cost?

ANALYSIS.—*If 1 apple cost 2 cents, 2 apples will cost twice 2 cents, which are 4 cents; and 1-half of an apple will cost 1-half of 2 cents, which is 1 cent; and 4 cents and 1 cent are 5 cents; therefore, if 1 apple cost 2 cents, 2 apples and 1-half will cost 5 cents.*

2. Two times 2 and 1-half of 2 are how many?

3. At 2 cents each, what will 3 pears and 1-half of a pear cost?

4. Three times 2 and 1-half of 2 are how many?

5. At $3 a yard, what will 3 yards and 1-third of a yard of cloth cost?

6. Three times 3 and 1-third of 3 are how many?

7. At $3 a barrel, what will 3 barrels and 2-thirds of a barrel of flour cost?

SUGGESTION.—To find 2-thirds of any number, first find 1-third, then multiply by 2.

8. Three times 3 and 2-thirds of 3 are how many?

9. At $3 a yard, what will 4 yards and 2-thirds of a yard of cloth cost?

ANALYSIS.—*If 1 yard cost $3, 4 yards will cost 4 times $3, which are $12; and 1-third of a yard will cost 1-third of $3, which is $1; and 2-thirds of a yard will cost twice as much as 1-third, that is, twice $1, which are $2; and $12 and $2 are $14. Ans. $14.*

10. If an orange cost 4 cents, what will 3 oranges and 1-fourth of an orange cost?

11. 3 times 4 and 2-fourths of 4 are how many?

12. 4 times 4 and 3-fourths of 4 are how many?

13. At $5 a barrel, what will 4 barrels and 2-fifths of a barrel of flour cost?

14. 4 times 5 and 2-fifths of 5 are how many?

15. 5 times 5 and 3-fifths of 5 are how many?

16. If a man spend $6 a week, how much will he spend in 4 weeks and 1-sixth of a week?

17. 4 times 6 and 2-sixths of 6 are how many?

18. 5 times 6 and 3-sixths of 6 are how many?

19. 6 times 6 and 4-sixths of 6 are how many?

20. 7 times 6 and 5-sixths of 6 are how many?

21. At 7 cents a yard, how much will 3 yards and 2-sevenths of a yard of tape cost?

22. 4 times 7 and 3-sevenths of 7 are how many?

23. 5 times 7 and 4-sevenths of 7 are how many?

24. 6 times 7 and 6-sevenths of 7 are how many?

25. If oranges are worth 8 cents each, how much are 3 oranges and 2-eighths of an orange worth?

26. 4 times 8 and 3-eighths of 8 are how many?

27. 5 times 8 and 4-eighths of 8 are how many?

28. 6 times 8 and 7-eighths of 8 are how many?

29. If a yard of muslin cost 9 cents, what will 2 yards and 2-ninths of a yard cost?

30. 3 times 9 and 5-ninths of 9 are how many?

31. 5 times 9 and 6-ninths of 9 are how many?

32. 7 times 9 and 8-ninths of 9 are how many?

33. If a pound of sugar cost 10 cents, what will 2 pounds and 2-tenths of a pound cost?

PARTS OF NUMBERS. 61

34. 2 times 10 and 3-tenths of 10 are how many?
35. 5 times 10 and 6-tenths of 10 are how many?
36. 6 times 10 and 8-tenths of 10 are how many?
37. 9 times 10 and 7-tenths of 10 are how many?
38. 10 times 9 and 8-ninths of 9 are how many?
39. 12 times 11 and 9-elevenths of 11 are how many?

SECTION IX.

LESSON I.

1. A boy bought 3 apples at 4 cents each: how much did they cost? He paid for them with oranges, at 6 cents each: how many oranges did it take?

ANALYSIS.—First. *3 apples cost 3 times the price of 1 apple, that is, 3 times 4 cents, or 12 cents.* Second. *It took as many oranges as 6 cents are contained times in 12 cents; 6 cents are contained in 12 cents 2 times. Ans. 2 oranges.*

2. A man bought 3 yards of cloth at $4 a yard: how many dollars did it cost? He paid for it with cider, at $2 a barrel: how many barrels did it take?

3. A boy bought 8 apples at 2 cents each: how much did they cost? He paid for them with pears, at 4 cents each: how many did it take?

4. Bought 9 marbles at 2 cents each; paid for them with tops, at 6 cents each: how many did it take?

5. Bought 10 yards of cloth at $2 a yard; paid for it with flour, at $4 a barrel: how many bar. did it take?

6. Bought 8 pints of cherries at 3 cents a pint; paid for them with apples, at 6 cents a dozen: how many dozen did it take?

7. How many barrels of flour, at $3 a barrel, must be given for 2 yards of cloth, at $7 a yard?

8. 5 times 3 are how many times 4? 6? 7? 8?
9. 4 times 4 are how many times 3? 5? 6? 7?
10. 5 times 4 are how many times 3? 6? 7? 8?
11. 2 times 11 are how many times 3? 4? 5? 6?
12. 5 times 5 are how many times 3? 4? 6? 7?
13. 8 times 4 are how many times 3? 5? 6? 7?
14. 7 times 5 are how many times 4? 6? 8? 9?
15. 6 times 7 are how many times 4? 5? 8? 9?
16. 6 times 9 are how many times 5? 7? 8? 10?
17. 8 times 7 are how many times 5? 6? 9? 10?
18. 6 times 10 are how many times 5? 7? 8? 10?
19. 9 times 7 are how many times 6? 8? 10? 11?
20. 8 times 8 are how many times 7? 9? 10? 11?

LESSON II.

1. Bought 4 boxes and 3-fifths of a box of raisins, at $5 a box: how much did they cost? Paid for them with flour, at $6 a barrel: how many did it take?

ANALYSIS.—*It will take as many barrels of flour as $6, the price of 1 barrel, are contained times in the cost of 4 boxes and 3-fifths of a box of raisins, at $5 a box.*

2. Bought 4 gallons and 4-sixths of a gallon of wine, for $6 a gallon, and paid for it with raisins, at $5 a box: how many boxes did it take?

3. Bought 5 kegs and 4-sevenths of a keg of tobacco, for $7 a keg, and paid for it with paper, at $6 a ream: how many reams did it take?

4. Five times 5 and 3-fifths of 5 are how many times 3? 4? 6? 7? 8? 9? 10?

5. Seven times 4 and 3-fourths of 4 are how many times 3? 5? 6? 8? 9? 10?

6. Five times 6 and 5-sixths of 6 are how many times 3? 4? 7? 8? 9? 10?

7. Seven times 5 and 3-fifths of 5 are how many times 4? 6? 8? 9? 10?

8. Five times 8 and 1-eighth of 8 are how many times 4? 6? 7? 9? 10?

9. Seven times 6 and 2-sixths of 6 are how many times 5? 8? 9? 10?

10. Nine times 5 and 2-fifths of 5 are how many times 6? 7? 8? 10?

11. Nine times 5 and 4-fifths of 5 are how many times 6? 7? 9? 10?

12. Six times 8 and 3-eighths of 8 are how many times 5? 7? 9? 10?

13. Seven times 7 and 4-sevenths of 7 are how many times 5? 6? 8? 9? 10?

14. Nine times 6 and 3-sixths of 6 are how many times 5? 7? 8? 10?

15. Eight times 7 and 3-sevenths of 7 are how many times 5? 6? 9? 10?

LESSON III.

1. Bought 4 apples at 3 cents each; paid for them with lemons at 6 cents each: how many did it take?

2. Bought 7 yards of tape, at 2 cents a yard: how many pears at 3 cents each, will it take to pay for it?

3. If 2 apples cost 4 cents, what cost 3 apples?

ANALYSIS.—*Three apples will cost 3 times as much as 1 apple, and 1 apple will cost 1-half as much as 2 apples, that is, 1-half of 4 cents. 1-half of 4 cents is 2 cents, and 3 times 2 cents are 6 cents. Ans. 6 cents.*

4. If 3 yards of cloth cost $9, what cost 4 yards?

5. If 3 oranges cost 15 cents, what cost 5 oranges?

6. If 4 barrels of flour cost $24, what cost 7 barrels?

7. If 2 kegs of lard cost $8, what cost 9 kegs? 11 kegs?

8. If 5 dozen eggs cost 35 cents, what will 3 dozen cost? What will 8 dozen cost?

9. If a man get $14 for 7 days' work, how much will he get for 9 days? For 3 days?

10. Bought 4 yards and 2-thirds of a yard of cloth, at $3 a yard, and paid for it with cheese, at $7 a hundred weight: how many hundred weight did it take?

11. Bought 4 pounds and 4-fifths of a pound of nails, at 5 cents a pound, and paid for them with eggs, at 3 cents a dozen: how many dozen did it take?

12. Bought 7 pounds and 5-sevenths of a pound of sugar, at 7 cents a pound, and paid for it with chickens, at 9 cents each: how many did it take?

13. Bought 9 pounds and 2-sevenths of a pound of sugar, at 7 cents a pound, and paid for it with eggs, at 6 cents a dozen: how many dozen did it take?

14. How many pounds, at 7 cents a pound, must you give for 8 and 2-ninth yards, at 9 cents a yard?

15. How many barrels, at $6 a barrel, must be given in exchange for 4 and 5-seventh yards, at $7 a yard?

16. Bought 5 pounds and 3-sevenths of a pound of butter, at 7 cents a pound, and paid for it with raisins, at 6 cents a pound: how many pounds did it take?

17. How many apples, at 2 cents each, can you buy for 6 cents? How many for 15 cents?

18. How many pears, at 3 for 7 cents, can you buy for 21 cents? For 35 cents?

19. If 6 pears are worth 2 oranges, how many oranges can you buy for 21 pears?

20. A man bought 15 yards of cloth, at the rate of 3 yards for $5: how many dollars did it cost?

21. If a man receive $5 for 4 days' work, how many dollars will he get for 12 days' work?

22. How many pears, at 3 for 5 cents, can you buy for 25 cents?

PARTS OF NUMBERS.   65

23. If 2 pears are worth 6 cents, how many pears must be given for 4 oranges, at 6 cents each?

24°. What will be the cost of 3 barrels and 9-elevenths of a barrel of sugar, at $11 a barrel?

25. What will be the cost of 3 boxes and 4-fifths of a box of butter, at $5 a box?

26. Find the cost of 4 and 5-sixth tons of hay, at $6 a ton.

27. How many dozen eggs, at 12 cents a dozen, will pay for 10 and 10-eleventh pounds of sugar, at 11 cents a pound?

SECTION X.

LESSON I.

1. If 1-half of an orange cost 3 cents, what will the whole orange cost?

ANALYSIS.—1 *orange will cost twice as much as* 1-*half of an orange, that is, twice* 3 *cents. But twice* 3 *cents are* 6 *cents; therefore, if* 1-*half of an orange cost* 3 *cents, the whole orange will cost* 6 *cents.*

2. Three is 1-half of what number?

ANALYSIS.—3 *is* 1-*half of twice* 3; *but twice* 3 *are* 6; *therefore,* 3 *is* 1-*half of* 6.

3. If 1-fourth of a barrel of cider cost $2, what will a barrel cost?

4. Two is 1-fourth of what number?

5. If a man can walk 2 miles in 1-third of an hour, how far can he walk in an hour?

6. Two is 1-third of what number?

7. The age of Charles is 1-third that of Thomas; Charles is 4 years old: how old is Thomas?

8. Four cents is the 1-third of what number of cents?

9. David has 5 marbles, which is only 1-fourth as many as Henry: how many has Henry?

ANALYSIS.—5 is 1-fourth of four times 5; but 4 times 5 are 20; therefore, 5 is 1-fourth of 20.

10. Six is 1-fourth of what number? 8 is 1-fourth of what number? 9 is 1-fourth of what?

11. Six is 1-third of what number? 7 is 1-third of what number? 9 is 1-third of what?

12. Three is 1-fifth of what number? 5 is 1-fifth of what number? 9 is 1-fifth of what?

13. Five is 1-sixth of what number? 7 is 1-sixth of what number?. 9 is 1-sixth of what?

14. Two is 1-seventh of what number? 5 is 1-seventh of what number? 8 is 1-seventh of what?

15. Five is 1-eighth of what number? 7 is 1-eighth of what number? 9 is 1-eighth of what?

16. Seven is 1-ninth of what number? 8 is 1-ninth of what number? 9 is 1-ninth of what?

LESSON II.

1. James had 4 apples, and gave his brother 1-half of them: how many did he give him?

2. If you divide 6 apples equally between 2 boys, what part of them must each have? *Ans.* 1-*half*.

3. If you divide 6 apples equally between 3 boys, what part of them must each have? *Ans.* 1-*third*.

4. If 3 yards of cloth cost $9, what part of $9 will 1 yard cost? What part of $9 will 2 yards cost?

5. If 4 oranges cost 12 cents, what part of 12 cents will 1 orange cost? What part of 12 cents will 2 cost? What part will 3 cost?

6. What is 1-fourth of 12? What are 2-fourths of 12? 3-fourths of 12?

PARTS OF NUMBERS.

7. If 5 barrels of flour cost $30, what part of $30 will 1 barrel cost? What part of $30 will 2 barrels cost? What part will 3 bar. cost? What part will 4 cost?

8. What is 1-fifth of 30? What are 2-fifths of 30? 3-fifths of 30? 4-fifths of 30?

9. If 1 apple cost 4 cents, what will 1-half of an apple cost? What will 3-halves cost?

10. If 2 oranges cost 8 cents, what part of 8 cents will 1 orange cost? 3 oranges? 5?

11. If 3 barrels of cider cost $12, what part of $12 will 1 barrel cost? What part of $12 will 2 barrels cost? 4 barrels? 5 barrels?

12. What is 1-third of 12? What are 2-thirds of 12? 4-thirds of 12? 5-thirds of 12?

13. What is 1-fourth of 24? 2-fourths? 3-fourths? 5-fourths? 6-fourths? 7-fourths? 9-fourths?

14. What is 1-fifth of 25? 2-fifths? 3-fifths? 4-fifths? 6-fifths? 7-fifths? 8-fifths? 9-fifths?

15. What is 1-sixth of 24? 2-sixths? 3-sixths? 4-sixths? 5-sixths? 6-sixths? 7-sixths?

16. What is 1-seventh of 56? 2-sevenths? 3-sevenths? 4-sevenths? 5-sevenths? 7-sevenths?

17. What is 1-eighth of 72? 2-eighths? 5-eighths? 8-eighths? 10-eighths? 11-eighths?

18. What is 1-ninth of 54? 3-ninths? 5-ninths? 7-ninths? 9-ninths? 11-ninths? 12-ninths?

19. What is 1-tenth of 60? 3-tenths? 7-tenths? 9-tenths? 11-tenths?

20. What are 2-thirds of 6?

ANALYSIS.—*Two-thirds of 6 are twice 1-third. One-third of 6 is 2, and twice 2 are 4; therefore, 2-thirds of 6 are 4.*

21. What are 2-thirds of 12?   3-fourths of 12?
22. What are 4-fifths of 20?   5-sixths of 30?
23. What is 1-ninth of 27?   1-fourth of 36?

24. What are 3-sevenths of 28? 2-fifths of 20?
25. What are 4-sixths of 24? 3-fourths of 20?
26. What are 5-ninths of 18? 6-sevenths of 21?
27. What are 6-sevenths of 49? 3-eighths of 24?
28. What are 5-eighths of 40? 4-ninths of 54?
29. What are 6-sevenths of 63? 7-eighths of 56?
30. What are 3-halves of 18? 4-thirds of 24?
31. What are 7-fourths of 12? 8-fifths of 30?
32. What are 8-sixths of 42? 9-sevenths of 63?
33. What are 9-eighths of 56? 10-ninths of 81?
34. What are 11-tenths of 50? 12-tenths of 40?

LESSON III.

1. A man having 12 bushels of grain, divided 5-sixths of it equally among 3 poor persons: how many bushels did each receive?

2. Five-sixths of 12 are how many times 3?

3. A boy having 25 apples, kept 1-fifth himself, and divided the other 4-fifths equally among 6 of his companions: how many did each receive?

4. 4-fifths of 25 are how many times 6?
5. 3-fourths of 24 are how many times 9?
6. 7-fourths of 24 are how many times 8?
7. 8-thirds of 18 are how many times 6?
8. 7-thirds of 27 are how many times 10?
9. 3-fifths of 60 are how many times 7?
10. 5-sixths of 54 are how many times 8?
11. 8-sixths of 48 are how many times 9?
12. 3-sevenths of 56 are how many times 9?
13. 9-sevenths of 63 are how many times 10?
14. 5-eighths of 64 are how many times 6?
15. 9-eighths of 40 are how many times 7?

PARTS OF NUMBERS.

16. 11-sevenths of 49 are how many times 8?
17. 3-ninths of 54 are how many times 7?
18. 10-ninths of 63 are how many times 8?
19. 8-ninths of 54 are how many times 5?
20. 9-sevenths of 42 are how many times 8?

LESSON IV.

1. James gave his brother 2 apples, which were 1-third of all he had: how many had he left?

2. Thomas gave his brother 5 cents, which were 1-fourth of all he had: how many had he?

3. If, in traveling, I walk 1-fifth of my journey in 2 hours, at that rate, in what time can I complete the remaining 4-fifths?

4. One pint is 1-eighth of a gallon: if a pint of wine cost 7 cents, what will a gallon cost?

5. A boy found a purse containing $12, and received 1-sixth of the money for returning it to the owner: how much did he receive?

6. If $42 be equally divided among a number of men, giving each man 1-sixth of the money, how many men would there be, and what would each receive?

7. Thomas had 28 marbles: he gave 1-fourth of them to James, and twice as many to William as to James: how many did each receive?

8. William had 24 apples: he gave 1-half to Thomas, and 1-third to James: how many did he give to both? How many had he left?

9. A boy had 12 cents: he spent 1-third of them for apples, and 1-fourth for cakes: how many had he left?

10. A little girl received 20 cents from her mother; her brother gave her 1-fifth as many as her mother, and her sister 1-half as many as her brother: how many did she then have?

2d Bk.   5

11. A boy having 40 cents gave 3-fifths of them for 2 arithmetics: what was the price of 1 arithmetic?

ANALYSIS.—*3-fifths are 3 times 1-fifth: 1-fifth of* 40 *cents is* 8 *cents, and 3-fifths are 3 times* 8 *cents, which are* 24 *cents. If 2 books cost* 24 *cents,* 1 *book will cost 1-half of* 24 *cents, which is* 12 *cents.* Ans. 12 *cents.*

12. James had 14 cents, and gave 4-sevenths of them to his sister: how many cents had he left?

13. John had 15 pears: he gave 1-third to Frank, and 3-fifths to Harry: how many had he left?

14. A man had 30 yards of cloth, and sold 2-fifths of it for $48: how much was that a yard?

15. John had 25 cents, and gave 3-fifths for peaches, at 2 cents each: how many did he buy?

16. A boy having 54 chestnuts, divided 5-ninths of them among 3 girls: how many did each receive?

17. A man had 28 barrels of flour, and sold 2-sevenths of them for $24: what was that a barrel?

18. Bought 7 yards of cloth for $42: I gave 3 yards for 9 barrels of cider: how much was it a barrel?

19. A man had $40, and lost 3-fifths of them: he expended the remainder in flour, at $4 a barrel: how many barrels did he buy?

20. I had 10 cents, and lost 1-fifth: spent the rest for apples, at 2 cents each: how many did I buy?

21. James had 48 cents: he gave 3-eighths to his brother, and spent the rest in chestnuts, at 5 cents a quart: how many quarts of chestnuts did he buy?

22. Thomas had 28 cents: he gave 1-fourth to his sister, and 3-sevenths to his brother, and with the remainder he bought 3 books: what did each cost?

SECTION XI.

LESSON I.

1. If 2-thirds of a melon cost 4 cents, what will 1-third cost?

ANALYSIS.—1-*third is* 1-*half of* 2-*thirds: if* 2-*thirds cost* 4 *cents,* 1-*third will cost* 1-*half of* 4 *cents, or* 2 *cents, Ans.*

2. Four is 2 times what number? *Ans.* Four is 2 times 1-half of 4, which is 2.

3. If 3-fourths of a yard of cloth cost $6, what will 1-fourth cost?

4. Six is 3 times what number?

5. If 2-thirds of a barrel of flour cost $8, what will 1-third cost?

6. Eight is 2 times what number?

7. If 3-fifths of a pound of butter cost 9 cents, what will 1-fifth cost?

8. Nine is 3 times what number?

9. If 4-fifths of a pound of coffee cost 16 cents, what will 1-fifth cost?

10. Sixteen is 4 times what number?

11. If 5-sixths of a gallon of wine cost 35 cents, what will 1-sixth cost?

12. If 6-tenths of a yard of cloth cost 30 cents, what will 1-tenth cost?

13. If 4-sevenths of a yard of muslin cost 28 cents, what will 1-seventh cost?

14. If 2-thirds of an orange cost 4 cents, what cost 1-third? If 1-third cost 2 cents, what cost the whole?

15. Four is 2-thirds of some number: what is 1-third of that number? 2 is 1-third of what number? Then 4 is 2-thirds of what number?

16. If 2-thirds of a yard of cloth cost $6, what cost 1-third of a yard? If 1-third cost $3, what cost a yard?

17. Six is 2-thirds of some number: what is 1-third of that number? 3 is 1-third of what number? Then 6 is 2-thirds of what number?

18. If 3-fourths of a barrel of flour cost $6, what will 1-fourth cost? If 1-fourth cost $2, what cost a barrel?

19. Six is 3-fourths of some number: what is 1-fourth of the same number? 2 is 1-fourth of what number? Then 6 is 3-fourths of what number?

20. If 2-fifths of a melon cost 8 cents, what cost 1-fifth? If 1-fifth cost 4 cents, what cost the whole?

21. Eight is 2-fifths of some number: what is 1-fifth of the same number? 4 is 1-fifth of what number? Then 8 is 2-fifths of what number?

22. If 3-fifths of a pound cost 9 cents, what cost 1-fifth? If 1-fifth cost 3 cents, what cost a pound?

23. Nine is 3-fifths of some number: what is 1-fifth of that number? 3 is 1-fifth of what number? Then 9 is 3-fifths of what number?

24. If 4-fifths of a pound cost 8 cents, what cost 1-fifth? If 1-fifth cost 2 cents, what cost a pound?

25. Eight is 4-fifths of some number: what is 1-fifth of that number? 2 is 1-fifth of what number? Then 8 is 4-fifths of what number?

26. If 3-fourths of a dozen eggs cost 9 cents, what cost 1-fourth? If 1-fourth cost 3 cents, what cost a dozen?

27. Nine is 3-fourths of some number, what is 1-fourth of that number? 3 is 1-fourth of what number? Then 9 is 3-fourths of what number?

28. If 2-thirds of a yard of tape cost 20 cents, what cost 1-third? If 1-third cost 10 cents, what cost a yard?

29. Twenty is 2-thirds of some number: what is 1-third of that number? 10 is 1-third of what number? Then 20 is 2-thirds of what number?

30. Six is 3-fourths of what number?

PARTS OF NUMBERS. 73

ANALYSIS.—1-*fourth* is 1-*third of* 3-*fourths:* if 3-*fourths* are 6, 1-*fourth* is 1-*third of* 6, *which is* 2, *and* 4-*fourths,* (*or the whole number,*) *are* 4 *times* 2, *which are* 8, *Ans.*

31. 9 is 3-sevenths of what number?
32. 10 is 2-sevenths of what number?
33. 14 is 7-eighths of what number?
34. 15 is 3-eighths of what number?
35. 16 is 4-ninths of what number?
36. 18 is 6-tenths of what number?
37. 20 is 5-fourths of what number?
38. 15 is 5-fourths of what number?
39º. 33 is 11-sixths of what number?
40º. 22 is 2-elevenths of what number?
41º. 81 is 9-thirds of what number?

LESSON II.

1. If 3-fourths of a pound of raisins cost 9 cents, how much will a pound cost? How many lemons, at 2 cents each, will pay for 1 pound of raisins?

ANALYSIS.—First. 1-*fourth will cost* 1-*third as much as* 3-*fourths, and* 4-*fourths, or a pound, will cost* 4 *times as much as* 1-*fourth; if* 3-*fourths cost* 9 *cents,* 1-*fourth will cost* 1-*third of* 9 *cents, or* 3 *cents, and* 4-*fourths will cost* 4 *times* 3 *cents, or* 12 *cents.*

Second. *At* 2 *cents each, it will require as many lemons as* 2 *cents are contained times in* 12 *cents;* 2 *cents are contained in* 12 *cents* 6 *times. Ans.* 6 *lemons.*

2. If 2-thirds of a pound of sugar cost 16 cents, how much will a pound cost? How many oranges, at 4 cents each, will pay for 1 pound of sugar?

3. If 7-eighths of a barrel of wine cost $42, how many bar. of cider, at $6 a bar., will pay for 1 bar. of wine?

4. If 3-fifths of a hogshead of sugar cost $24, how many barrels of flour, at $4 a barrel, will pay for 1 hogshead of sugar?

5. Sold a horse for $25, which was 5-eighths of his cost: how much did he cost me? I paid for him with cloth, at $6 a yard: how many yards did I give?

6. Thirty is 5-eighths of how many times 5?

Ans. Of as many times 5, *as* 5 *is contained times in the number of which* 1-*fifth of* 30 *is* 1-*eighth.*

ANALYSIS.—*If* 30 *is* 5-*eighths,* 1-*fifth of* 30, *which is* 6, *is* 1-*eighth, and the number is* 8 *times* 6, *which are* 48; 48 *is* 9 *times* 5 *and* 3-*fifths of* 5.

7. 12 is 4-sevenths of how many times 5?
8. 18 is 3-eighths of how many times 9?
9. 16 is 2-sevenths of how many times 9?
10. 36 is 4-sevenths of how many times 8?
11. 45 is 5-ninths of how many times 7?
12. 24 is 4-thirds of how many times 5?
13. 72 is 8-fifths of how many times 7?
14. 81 is 9-fourths of how many times 3?
15. 50 is 10-sevenths of how many times 4?
16. 63 is 7-sixths of how many times 5?

LESSON III.

1. James gave his brother 4 marbles, which were 2-thirds of all he had: how many marbles had he?

2. Thomas sold a knife for 15 cents, which were 3 fifths of its cost: how much did it cost?

3. William lost 6 marbles, which were 3-eighths of all he had: how many had he?

4. I sold a horse for $42, which were 6-fifths of his cost: how many dollars did I gain?

PARTS OF NUMBERS. 75

5. A grocer sold a lot of flour for $40, which were 5-fourths of the cost: what was the cost?

6. Sold a horse for $56, which were 8-fifths of the cost: paid with flour, at $4 a barrel: how many barrels did I give?

7. A man sold a watch for $28, which were 4-thirds of its cost: how much did it cost?

8. A man purchased a horse: after paying 3-fifths of the price, he owed $20: what was the cost?

9. Alexander sold a book for 25 cents, and lost 2-sevenths of the cost: what was the cost?

10. If a boy can earn 32 cents in a whole day, how much can he earn in 5-eighths of a day?

11. If 7-ninths of a cask of wine cost $42, how much flour, at $8 a barrel, will pay for 1 cask of wine?

12. If there are 10 links in 5-ninths of a chain, how many links are there in the whole chain?

13. In an orchard there are 12 cherry-trees: the remaining 5-sevenths of the orchard are apple-trees: how many apple-trees are there?

14. There is a pole, 4-fifths of which is under water, and 6 feet out of water: how long is the pole?

15. There is a pole, 3-fifths of which is in the earth, and 12 feet in the air: how long is the pole?

16. One-fifth of a pole is in the mud, 2-fifths in the water, and 14 feet in the air: how long is the pole?

17. The age of Joseph is 25 years, which is 5-eighths of the age of his father: the father's age is 10 times that of his youngest son: what is the age of the father? what is the age of his youngest son?

18. A man sold a horse for $45, which were 5-thirds of the cost: how much did he gain?

19. A man paid $24 for a watch, and sold it for 7-fourths of the cost: he was paid in cloth, at $5 a yard: how many yards did he receive?

20. A watchmaker sold a watch for $18, and lost 2-fifths of its value: how much did he lose?

ANALYSIS.—*Since he lost 2-fifths, he sold it for 3-fifths of its value; then 3-fifths are $18, and 1-fifth is 1-third of $18, or $6; 2-fifths are 2 times $6, or $12, Ans.*

21. A watchmaker sold a watch for $45, and gained 2-sevenths of the cost: what was the cost?

22. A boy spent 3-sevenths of his money, and had 12 cents left: how much had he?

SECTION XII.

LESSON I.

1. If you have 12 cents, and 3-fourths of your money equal 1-third of mine, how many cents have I?

ANALYSIS.—*1-fourth of 12 is 3, and 3-fourths are 3 times 3, or 9; 9 is 1-third of 3 times 9: 3 times 9 are 27, Ans.*

2. 2-thirds of 9 are 1-fifth of what number?
3. 4-fifths of 10 are 1-half of what number?
4. 3-sevenths of 14 are 1-sixth of what number?
5. 5-sixths of 12 are 1-fourth of what number?
6. 3-eighths of 16 are 1-fifth of what number?
7. 2-fifths of 20 are how many thirds of 15?
8. 4-sevenths of 21 are how many fifths of 25?
9. 5-sixths of 30 are how many fourths of 32?
10. 7-ninths of 45 are how many sevenths of 21?
11. 3-fourths of 36 are how many thirds of 18?
12. 5-eighths of 40 are how many ninths of 36?

13. Divide 1-fourth of 32 by 2-thirds of 9. Three-fourths of 28 by 2-thirds of 12.

PARTS OF NUMBERS. 77

14. Divide 3-fifths of 40 by 3-sevenths of 14. Five-sevenths of 42 by 3-eighths of 24.

15. Divide the 7-ninths of 54 by 5-sixths of 12.

16. Nine-elevenths of 88 are how many times 5-ninths of 18?

LESSON II.

1. William says to Frank, "Your age is 15 years, and 4-fifths of your age are 6-sevenths of mine: what is my age?"

ANALYSIS.—1-*fifth of* 15 *is* 3, *and* 4-*fifths are* 4 *times* 3, *which are* 12; *if* 12 *is* 6-*sevenths,* 1-*sixth of* 12, *which is* 2, *is* 1-*seventh, and if* 2 *is* 1-*seventh, the number is* 7 *times* 2, *which are* 14. *Ans.* 14 *years.*

2. 3-fourths of 8 are 2-thirds of what number?
3. 4-fifths of 20 are 8-ninths of what number?
4. 3-fifths of 20 are 4-fifths of what number?
5. 5-sixths of 36 are 10-thirds of what number?
6. 6-sevenths of 28 are 3-fifths of what number?
7. 5-eighths of 32 are 4-sevenths of what number?
8. 8-ninths of 45 are 10-elevenths of what number?
9. Five-sevenths of 21 are 3-fourths of 5 times what?
10. Four-ninths of 36 are 8-sevenths of 7 times what?
11. Three-tenths of 40 are 6-fifths of 10 times what?
12. Seven-eighths of 64 are 7-ninths of 12 times what?
13. Three-fourths of 16 are 2-thirds of how many times 1-fifth of 20?

ANALYSIS.—1-*fourth of* 16 *is* 4, *and* 3-*fourths are* 3 *times* 4, *which are* 12; *if* 12 *is* 2-*thirds,* 1-*half of* 12, *which is* 6, *is* 1-*third, and* 6 *is* 1-*third of* 3 *times* 6, *which are* 18; 1-*fifth of* 20 *is* 4, *and* 18 *is* 4 *times* 4 *and* 2-*fourths of* 4. *Ans.* 4 *and* 2-*fourths times.*

14. Four-fifths of 20 are 8-sevenths of how many times 1-third of 15?

15. Six-sevenths of 21 are 2-fifths of how many times 1-fifth of 30?

16. Four-ninths of 27 are 2-thirds of how many times 1-seventh of 35?

17. Five-sevenths of 42 are 5-ninths of how many times 1-eighth of 56?

18. Three-fourths of 24 are 9-tenths of how many times 2-fifths of 10?

19. Five-sixths of 12 are 2-sevenths of how many times 3-fourths of 8?

20. Five-eighths of 32 are 4-ninths of how many times that number of which 2 is 1-third?

21. Six-sevenths of 28 are 8-elevenths of how many times that number of which 4 is 2-fifths?

22. Eight-ninths of 45 are 10-thirds of how many times that number of which 14 is 7-fourths?

23. Seven-tenths of 50 are 5-ninths of how many times that number of which 8 is 4-fifths?

24. Nine-sevenths of 56 are 12-fifths of how many times the number of which 6 is 6-sevenths?

25. One-third of a certain number is 2 more than 1-half of 12: what is the number?

26. One-fourth of a certain number is 3 less than 1-fifth of 30: what is the number?

27. Two-fifths of 20 is 6 less than how many thirds of 21?

28. Three-fourths of 24 is 6 more than 2-thirds of what number?

29. Five-sixths of 30, increased by 4, is 1 less than 3-fourths of some number: what is that number?

30. Three-fifths of 40 is just 3 less than 9-sevenths of how many times 7?

SECTION XIII.—REVIEW.

LESSON I.

1. William had 23 cents: Thomas gave him 8 cents more, George 6, James 5, and David 7: he gave 15 cents for a book: how many cents had he left?

2. A grocer paid $12 for sugar, $9 for coffee, $5 for tea, $7 for flour, and had $10 left: how many dollars had he at first?

3. A boy had 11 cents: his father gives him 9 cents, his mother 6, and his sister enough to make 34: how many cents did his sister give him?

4. Five men bought a horse for $42: the first gave $13, the second 7, the third 5, and the fourth 9: how many dollars did the fifth give?

5. A man purchased 8 sheep at $4 a head, 5 barrels of flour at $3 a barrel, 4 yards of cloth at $3 a yard, and 5 ounces of opium at $1 an ounce: how much did he spend?

6. A boy lost 25 cents: after finding 15 cents, he had 25 cents: how many cents had he at first?

7. A man owed a debt of $28, and paid all but $9: how much did he pay?

8. Borrowed $56: at one time paid $23; at another, all but $7: how much did I pay the last time?

9. James borrowed 37 cents: at one time he paid 5, at another 8, and the third time, all but 15.: how many cents did he pay the third time?

10. A farmer sold 2 cows at $9 each, and 5 hogs at $3 each, and received in payment 3 sheep at $3 each, and the rest in money: how much money did he receive?

11. A farmer sold 12 barrels of cider at $3 a barrel: he then purchased 5 barrels of salt at $3 a barrel, and some sugar, for $8: how many $'s had he left?

12. A merchant purchased 13 hats at $4 each, 5 pairs of shoes at $2 a pair, and an umbrella, for $7: what must he sell the whole for, to gain $9?

13. If 1 barrel of flour cost $3, what cost 7 barrels?

14. If 2 pounds cost 16 cents, what cost 5 pounds?

15. If 3 barrels of cider cost $12, what cost 4 barrels?

16. If 4 yards of cloth cost $28, what cost 7 yards?

17. If 3 pounds cost 27 cents, what cost 8 pounds?

18. If 5 barrels of flour cost $35, what cost 8 barrels?

19. If 7 apples cost 28 cents, what cost 3 apples?

20. If 8 oranges are worth 24 apples, how many apples are 3 oranges worth?

21. If 4 pounds of cheese cost 36 cents, what will 3 pounds cost?

22. If 8 yards of cloth cost $56, what cost 7 yards?

23. What sum of money must be divided among 11 men that each shall receive $9?

24. A walks 5 miles, while B walks 3: when A has gone 35 miles, how far has B gone?

ANALYSIS.—*In* 35 *there are* 7 *fives, and B walks as many threes as A walks fives; that is,* 7 *times* 3, *or* 21 *miles.*

25. Joseph and his father are husking corn: the father can husk 7 rows while Joseph husks 3: how many rows will Joseph husk while his father husks 42?

26. Charles can earn $9 while Mary earns $4: how many $'s will Charles earn while Mary earns 28?

27. If 6 horses eat 12 bushels of oats in a week, how many will 10 horses eat in the same time?

28. If 5 horses eat 16 bushels in 2 weeks, how long would it take them to eat 56 bushels?

29. If 6 apples are worth 18 cents, how many apples must be given for 5 oranges worth 6 cents each?

30. How many horses can eat in 9 days, the same amount of hay that 12 horses eat in 6 days?

31°. If 5 men earn $30 in 3 days, how much will 2 men earn in the same time? How much will 2 men earn in 1 day?

LESSON II.

1. How many times 6 in 3-fifths of 40?
2. How many times 3-sevenths of 14 in 54?
3. How many times 5-sixths of 12, in 4-ninths of 72?
4. How many times 3-fifths of 20, in twice that number of which 14 is 7-ninths?
5. If 3-eighths of a tun of hay cost $9, what will 5-sixths of a tun cost?
6. If a man can earn $7 in 1-fifth of a month, how many dollars can he earn in 1 month?
7. If a man can earn $42 in 1 month, how many dollars can he earn in 1-sixth of a month?
8. If a hogshead of sugar is worth $96, what are 7-eighths worth?
9. Five men paid $20 for a horse: what part of $20 did 2 men pay?
10. If $7 will buy 56 yards of muslin, how many yards will $4 buy?
11. If 3 men can do a job of work in 16 days, in how many days can 4 men do it?
12. If 3 men spend $12 in 1 week, at the same rate, how many would 2 men spend in 6 weeks?
13. If 6 men do a piece of work in 7 days, in how many days can 3 men do it?
14°. If 5 men do a piece of work in 8 days, in how many days can 4 men do a job twice as large?
15°. If 6 men do a job in 5 days, in how many days can 2 men do a job half as large?
16. James had 16 apples: he kept 1-fourth himself and divided the remainder equally among 3 of his companions: how many did each receive?

17. Three-fourths of 24, increased by 2-thirds of 12, are equal to how many?

18. Five-sixths of 24, diminished by 3-fourths of 20, equal how many?

19. One-half, and 2-thirds, and 3-fourths of 12 are how many?

20. Two-thirds of 12, less 1-half of 12, are 2-fifths of what number?

21°. From 10 take 3-fifths of itself; add to the remainder its 1-half: what is the result?

22. Thomas had 28 cents: he gave 2-sevenths to his sister, and 2-fifths of the remainder to his brother: how many more did he give, than he had left?

23. James had 35 marbles: he gave to Thomas 3-sevenths, to Charles 2-fifths: to which did he give the most, and how many? What number had he left?

24. Thomas had $28: he kept 2-sevenths, and divided the remainder equally among his 4 brothers: how many dollars did each receive?

25. A grocer had 14 barrels of flour: he sold 4-sevenths at $3 a barrel, and the remainder at $5 a barrel: what did it amount to?

26. Bought 15 yards of cloth, at $2 a yard: I sold 1-third at $4 a yard, 2-fifths at $3 a yard, and the remainder at $5 a yard: how much did I gain?

27. Bought 10 yards of cloth for $90, and sold 2-fifths of it for $40: how much a yard did I gain?

28. Two men travel the same direction: A is 40 miles ahead of B; but B travels 23 miles a day, and A 18: in how many days will B overtake A?

29. A hare is 90 yards in advance of a hound: the hound goes 10 yards in a minute, and the hare 7: in how many minutes will the hound overtake the hare? How far will each run?

30°. If a hound runs 7 rods while a hare runs 4, how far will the hare run while the hound runs 35 rods?

31. One man is pursuing another, who is 4 miles in advance: he travels 3 miles in pursuit, while the other advances 1: how much does he gain in going 3 miles? How far must he travel to overtake him?

32°. C and D travel in the same direction: C is 15 miles ahead of D; but D travels 5 miles an hour, and C only 2: in how many hours will D overtake C? How far will D have traveled?

33. A cistern containing 24 gallons, is filled by a pipe at the rate of 8 gallons an hour, and emptied by a pipe at the rate of 5 gallons an hour: if both pipes are open, how many gallons will remain in the cistern each hour? How long will the cistern be in filling?

34°. A cistern containing 36 gallons has 2 pipes; by the first it receives 6 gallons an hour, and by the second it discharges 9 gallons an hour: if both pipes are left open, how long will it take to empty the cistern?

35. My pants cost $8, which were 2-fifths the cost of my coat; my vest cost 1-half as much as my pants: what was the cost of the whole?

LESSON III.

1. What are the divisors of a number called? *Ans. Its factors.*

2. What is the divisor of a number? *Ans. Any number which is contained in it an exact number of times; that is, without a remainder.*

3. What is a multiple of a number? *Ans. Any product of which that number is a factor.*

4. Twelve can be divided by 2, 3, 4, and 6: what are the factors of 12? of 24?

5. What 2 numbers multiplied together, will make 28? What are the factors of 28?

6. What numbers will divide 22? What are the factors of 22?

7. Six is a factor of 24; what number is a multiple of 6? of 12?

8. Seven is a factor of 21? What number is a multiple of 7? of 3?

9. What three numbers are multiples of 7? of 8? of 9? of 10?

10. What numbers will divide 18? 27? 33? 36? 39? 48? 49? 63? 64? 65? 72? 75?

11. What number will divide both 10 and 15? 12 and 16? 20 and 32? 64 and 80? 72 and 96?

12. What is the greatest number less than 12, that will divide 12? Less than 16 that will divide 16? Less than 20 that will divide 20?

13. What is the greatest number that will divide both 6 and 10? 12 and 18? 20 and 32? 24 and 36? 42 and 48? 56 and 72?

14. What numbers can be divided by 9? 12? 8? 7? 16? 14? 15?

15. What is the least number that can be divided by 2 and 3? 4 and 7? 2 and 9? 7 and 21? 10 and 12? 12 and 20?

The answers to the 13th question are called the *Greatest Common Divisors;* to the 15th question, the *Least Common Multiples* or *Dividends.*

SECTION XIV.—FRACTIONS.

LESSON I.

A single or whole thing of any kind, is called a *unit* or *one;* as 1 yard, 1 dollar, 1 mile, 1, &c.

1. The above line represents a yard of tape: it is called 1 yard.

2. If you divide it into *two* equal parts, one of the parts is called *one-half of a yard.*

one-half.   *one-half.*

One part is represented thus, ½, and is read *one-half.*

3. If the yard of tape is divided into *three* equal parts, *one* of the parts is called *one-third; two* parts are called *two-thirds.*

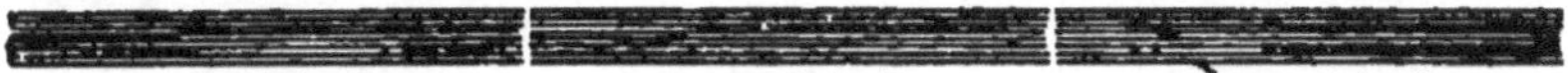

one-third.   *one-third.*   *one-third.*

One part is represented thus, ⅓, and is read *one-third;* 2 parts are represented thus, ⅔, read *two-thirds.*

4. If the yard of tape is divided into *four* equal parts, *one* of the parts is called *one-fourth; two* parts *two-fourths;* 3 parts *three-fourths.*

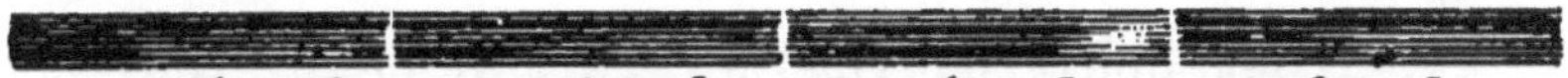

one-fourth.   *one-fourth.*   *one-fourth.*   *one-fourth.*

One part is represented thus, ¼, and is read *one-fourth;* 2 parts are represented thus, 2/4, read *two-fourths;* 3 parts thus, ¾, read *three-fourths.*

5. If the yard of tape is divided into *five* equal parts, one of the parts is called *one-fifth; two* parts *two-fifths; three* parts *three-fifths.*

one-fifth.   *one-fifth.*   *one-fifth.*   *one-fifth.*   *one-fifth.*

One part is represented thus, ⅕, and is read *one-fifth;* 2 parts thus, ⅖, read *two-fifths;* 3 parts thus, ⅗, read *three-fifths;* 4 parts thus, ⅘, read *four-fifths.*

A FRACTION is a part of a unit, or one. In every fraction the unit is divided into *equal* parts. The lower number shows into how many equal parts the unit is divided: the upper number, how many parts are taken.

The lower number is called the DENOMINATOR, because it *denominates* or names the parts: the upper number the NUMERATOR, because it *numbers* the parts.

In reading fractions, that is, in expressing them by words, read the Numerator first; the Denominator last.

6. If you divide a yard into 6 equal parts, what figures will express 2 parts? What 5 parts?

7. If you take away 2 of the 6 parts, what fraction will express the remainder?

8. How can you express in figures 3-sevenths of 1?

9. In the fraction $\frac{2}{5}$, into how many parts is the unit (one) divided, and how many are represented?

10. Read the following Fractions:

$\frac{1}{3}, \frac{1}{4}, \frac{1}{5}, \frac{1}{6}, \frac{1}{7}, \frac{1}{8}, \frac{1}{9}, \frac{1}{10}, \frac{1}{11}, \frac{1}{12}, \frac{1}{13}, \frac{1}{14}, \frac{1}{15}, \frac{1}{16}, \frac{1}{17},$

$\frac{2}{3}, \frac{3}{4}, \frac{4}{5}, \frac{5}{6}, \frac{4}{7}, \frac{7}{8}, \frac{5}{9}, \frac{3}{10}, \frac{9}{11}, \frac{10}{12}, \frac{9}{13}, \frac{13}{14}, \frac{15}{16}, \frac{19}{25}, \frac{32}{40}.$

11. What does the fraction $\frac{7}{8}$ represent? Into how many equal parts is the unit (one) divided?

12. What do you understand by $\frac{9}{11}$? $\frac{12}{15}$? $\frac{20}{31}$?

13. Divide an apple into 10 equal parts, and take away 3 parts: what fraction will express the remainder?

14. What fraction will represent 1-half of 1? 1-half of 3? 1-third of 5?

15. Which is the greater fraction, $\frac{1}{2}$ or $\frac{1}{3}$? $\frac{1}{5}$ or $\frac{1}{4}$? $\frac{2}{4}$ or $\frac{1}{2}$? $\frac{2}{3}$ or $\frac{3}{4}$?

16. What is a fraction? What does the denominator of a fraction show? What does the numerator show?

17. How many apples in 2-halves of an apple? In 3-thirds? In 4-fourths? In 5-fifths?

18. When the numerator of a fraction is equal to the denominator, as $\frac{2}{2}, \frac{3}{3}, \frac{4}{4}$, &c., what is its value? *Ans.* 1.

19. Are 3-halves of an apple more or less than a whole apple? When the numerator is greater than the denominator, is the value of the fraction greater or less than 1?

FRACTIONS. 87

20. If the denominators of fractions be increased, do the fractions become greater or less? Why?

21. If the numerators be increased, do the fractions become greater or less? Why?

22. Such fractions as $\frac{2}{2}$, $\frac{4}{3}$, $\frac{6}{4}$, &c., being either equal to, or greater than 1, are called *improper fractions;* those less than 1, as $\frac{1}{2}$, $\frac{2}{3}$, &c., *proper fractions.*

23. Whole numbers and fractions joined together, as $1\frac{1}{2}$, $2\frac{2}{3}$, $5\frac{3}{4}$, $6\frac{4}{7}$, $10\frac{5}{8}$, &c., are called *mixed numbers.*

SECTION XV.

LESSON I.

1. How do you find the number of halves in any number? Why?

2. In 1, how many halves? In $1\frac{1}{2}$? In 2? In $2\frac{1}{2}$? In 3? In $3\frac{1}{2}$? In 4? $4\frac{1}{2}$? 5? 6? 7? $7\frac{1}{2}$?

3. How do you find how many thirds there are in any number? Why?

4. How many thirds in 1? In $1\frac{1}{3}$? In $1\frac{2}{3}$? In 2? In $2\frac{1}{3}$? In $2\frac{2}{3}$? 3? $3\frac{1}{3}$? $3\frac{2}{3}$? 4? $4\frac{1}{3}$? 5? 6?

5. How do you find how many fourths there are in any number? Why?

6. How many fourths in 1 apple? In 1 and 1-fourth? In 1 and 2-fourths? In 1 and 3-fourths? In 2? In 2 and 3-fourths? In 3? In 4 and 3-fourths?

7. How find how many fifths are in any number?

8. How many fifths in 1? In $1\frac{1}{5}$? In $1\frac{2}{5}$? In $1\frac{3}{5}$? In $1\frac{4}{5}$? In 2? In $2\frac{1}{5}$? In $2\frac{2}{5}$? In $2\frac{3}{5}$? In $2\frac{4}{5}$? In 3? In $3\frac{2}{5}$? In $3\frac{4}{5}$? In $4\frac{2}{5}$? In $5\frac{1}{5}$?

9. In 4 and 2-sevenths how many sevenths?

10. In 5 and 3-eighths how many eighths?

11. In 4 and 7-ninths how many ninths?

12. In 5 and 3-tenths   how many tenths?
13. In 12 and 1-third   how many thirds?
14. In 11 and 1-half   how many halves?
15. In 6 and 3-fifths   how many fifths?
16. In 7 and 3-fourths   how many fourths?
17. In 6 and 1-sixth   how many sixths?
18. In 5 and 7-eighths   how many eighths?
19. In 11 and 4-sevenths how many sevenths?
20. In 7 and 9-tenths   how many tenths?

LESSON II.

1. James's brother gave him 3-halves of an apple: how many apples did he give him? *Ans. One apple and 1-half.*

2. In 3 halves how many ones?

ANAL.—*Since 2-halves make* 1, *in 3-halves are as many ones as 2-halves are contained times in 3-halves; 2-halves in 3-halves, one and one-half times. Ans. 1 and 1-half.*

3. John's father gave him 4 half-dollars: how many dollars did he give him?

4. How many ones in 4-halves? in 5-halves? in 6-halves? in 7-halves? in 8-halves? in 9-halves?

5. How do you find how many ones there are in any number of halves?

6. If you divide an orange into 3 equal parts, what is 1 part called? How many thirds in 1 orange?

7. If 1 orange make 3-thirds, how many oranges would there be in 3-thirds of an orange? in 4-thirds of an orange? in 5-thirds? in 6-thirds?

8. How many ones in 3-thirds? in 4-thirds? in 5-thirds? in 6-thirds? in 7-thirds? in 8-thirds?

9. How do you find how many ones there are in any number of thirds?

FRACTIONS.

10. In 4-fourths of a dollar how many dollars?
    In 5-fourths? In 6-fourths?
    In 7-fourths? In 8-fourths?
    In 9-fourths? In 11-fourths?
    In 13-fourths? In 15-fourths?
    In 18-fourths? In 19-fourths?

11. How many ones in 4-fourths? in 5-fourths? in 6-fourths? in 7-fourths? in 8 fourths? in 9-fourths? 11-fourths? 13-fourths? 15-fourths? 19-fourths?

12. How do you find how many ones there are in any number of fourths?

13. In 5-fifths of an orange, how many oranges?
    In 6-fifths? In 7-fifths?
    In 8-fifths? In 9-fifths?
    In 10-fifths? In 11-fifths?
    In 13-fifths? In 15-fifths?
    In 17-fifths? In 18-fifths?

14. How many ones in 5-fifths? 6-fifths? 7-fifths? 8-fifths? 9-fifths? 10-fifths? 11-fifths? 13-fifths?

15. How do you find how many times 1 there are in any number of fifths?

16. How many times 1, that is, how many whole ones, in 23-sixths? *Ans. Three and 5-sixths.*

17. In 30-sevenths are how many times 1?
18. In 35-sevenths?   22. In 23-halves?
19. In 46-ninths?   23. In 33-fifths?
20. In 53-tenths?   24. In 31-fourths?
21. In 37-thirds?   25. In 37-sixths?

26. In 47-eighths are how many whole ones?
27. In 81-sevenths?   30. In 75-tenths?
28. In 79-tenths?   31. In 89-elevenths?
29. In 53-ninths?   32. In 93-twelfths?

SECTION XVI.

LESSON I.

ILLUSTRATION.—The first line represents a yard of tape divided into two equal parts; the second, a yard of tape divided into four equal parts.

one-half.   *one-half.*

one-fourth.   one-fourth.   one-fourth.   one-fourth.

One-half ($\frac{1}{2}$), is equal to two-fourths ($\frac{2}{4}$).

1. If I give to Mary 1-half of an orange, and to Jane 1-fourth, how much more will Mary have than Jane? what part will be left?

2. How much greater is 1-half than 1-fourth? How much are 1-half and 1-fourth?

3. James divided a melon, giving to his sister 1-half, and to his brother 1-fourth: what part did he give away?

4. Thomas gave 3-fourths of a dollar for a geography, and 1-half of a dollar for an arithmetic and slate: how much did he give for both?

5. How much are 1-half and 3-fourths?

ILLUSTRATION.—The first line represents a yard of ribbon divided into 2 equal parts; the second, a yard divided into six equal parts.

one-half.   *one-half.*

1-*sixth*.   1-*sixth*.   1-*sixth*.   1-*sixth*.   1-*sixth*.   1-*sixth*.

6. One-half is how many sixths? 1-third is how many sixths? 2-thirds are how many sixths?

FRACTIONS. 91

7. James received 1-half of an orange, and Charles 1-third: how much more had James than Charles?

8. How much are 1-half and 1-third?

9. If 1-third is 2-sixths, how many sixths are there in 2-thirds?

10. A yard of flannel costs half a dollar: a yard of cloth 2-thirds of a dollar: how much do both cost?

11. James bought 2 melons; he gave to Lucy half of the first; to Jane 2-thirds of the second: what part of a melon had Jane more than Lucy?

12. How much greater are 2-thirds than 1-half?

13. After taking away 1-half and 1-third of an apple, what part will be left?

14. I wish to divide an orange, and give to Mary 1-half, to Jane 1-fourth, and to William 1-eighth: how must I divide it? how many eighths will each have?

15. Two-fourths are how many eighths? 3-fourths are how many eighths?

16. One-fifth is how many tenths? 2-fifths are how many tenths? 3-fifths? 4-fifths?

17. Thomas wishes to divide an orange, and give Ann 1-half, and Lucy 2-fifths: how must he divide it? what part will he have left?

18. How much are 1-half and 2-fifths?

19. One-third is how many twelfths? 1-fourth is how many twelfths? How many twelfths in 2-thirds? in 2-fourths? in 3-fourths?

20. A farmer sows 1-half of a field in wheat, 1-third in rye; the rest in barley: how many twelfths are in wheat? how many in rye? how many in barley?

21. How many twelfths in 1-fourth? in 1-fourth and 1-third?

22. David bought a pound of figs: he gave 1-third of them to his mother, 1-fourth to his sister, and 1-sixth to his brother: what part had he left?

23. How many eighths are in 1-half? in 3-fourths?

24. How many tenths are in 1-half? in 1-fifth? in 2-fifths? in 3-fifths?

25. How many fifteenths in 1-third? in 1-fifth? in 3-fifths? in 4-fifths?

26. How many twentieths in 1-half? in 1-fourth? in 1-fifth? in 3-fourths? in 3-fifths?

27. Reduce $\frac{2}{3}$ and $\frac{3}{4}$ to twelfths: $\frac{1}{3}$ and $\frac{2}{5}$ to fifteenths: also, $\frac{1}{2}$ and $\frac{4}{5}$ to tenths.

28. Reduce $\frac{2}{5}$ and $\frac{3}{4}$ to twentieths: $\frac{1}{2}$, $\frac{1}{4}$, and $\frac{1}{5}$ to twentieths.

29. Reduce $\frac{1}{2}$ and $\frac{4}{7}$ to fourteenths: $\frac{1}{2}$ and $\frac{5}{9}$ to eighteenths.

30. Reduce $\frac{2}{3}$, $\frac{3}{4}$, and $\frac{5}{8}$ to twenty-fourths: $\frac{5}{8}$ and $\frac{3}{10}$ to fortieths.

REMARK.—When two or more fractions have the *same* denominator, they are said to have a *common* denominator; thus, $\frac{2}{8}$ and $\frac{3}{8}$ have 6 for a common denominator.

When the denominators of two fractions are not the same, a common denominator may be found by multiplying the denominators together.

Find the Common Denominator

31. Of $\frac{1}{2}$ and $\frac{2}{3}$. $\frac{3}{4}$ and $\frac{1}{3}$. $\frac{2}{3}$ and $\frac{3}{5}$.
32. Of $\frac{3}{8}$ and $\frac{2}{3}$. $\frac{2}{5}$ and $\frac{3}{4}$. $\frac{1}{5}$ and $\frac{1}{6}$.
33. Of $\frac{3}{5}$ and $\frac{3}{4}$. $\frac{3}{4}$ and $\frac{5}{6}$. $\frac{1}{3}$ and $\frac{1}{7}$.
34. Of $\frac{2}{3}$ and $\frac{5}{8}$. $\frac{3}{4}$ and $\frac{5}{7}$. $\frac{2}{3}$ and $\frac{5}{9}$.
35. Of $\frac{2}{5}$ and $\frac{5}{7}$. $\frac{1}{5}$ and $\frac{3}{8}$. $\frac{1}{5}$ and $\frac{1}{9}$.
36. Of $\frac{1}{3}$ and $\frac{3}{10}$. $\frac{3}{4}$ and $\frac{2}{11}$. $\frac{2}{5}$ and $\frac{5}{12}$.
37. Of $\frac{1}{2}$ and $\frac{4}{7}$. $\frac{2}{3}$ and $\frac{7}{10}$. $\frac{3}{5}$ and $\frac{5}{6}$.
38. Of $\frac{3}{4}$ and $\frac{4}{9}$. $\frac{4}{5}$ and $\frac{5}{6}$. $\frac{2}{3}$ and $\frac{7}{8}$.

39°. Reduce $\frac{1}{2}$, $\frac{3}{4}$, $\frac{7}{8}$, and $\frac{15}{16}$ to thirty-seconds.

40°. Reduce $\frac{1}{3}$, $\frac{8}{9}$, $\frac{26}{27}$, and $\frac{53}{54}$ to fifty-fourths.

FRACTIONS. 93

41°. Reduce $\frac{2}{3}$, $\frac{5}{6}$, $\frac{3}{8}$, $\frac{7}{12}$, and $\frac{9}{16}$ to forty-eighths.

42°. Reduce $\frac{1}{3}$, $\frac{1}{4}$, $\frac{1}{6}$, $\frac{1}{8}$, $\frac{1}{9}$, $\frac{1}{12}$, and $\frac{1}{18}$ to seventy-seconds.

LESSON II.

ILLUSTRATION.—The first line represents a yard of ribbon divided into 2 equal parts; the second line, a yard divided into 4 equal parts.

one-half.   *one-half.*

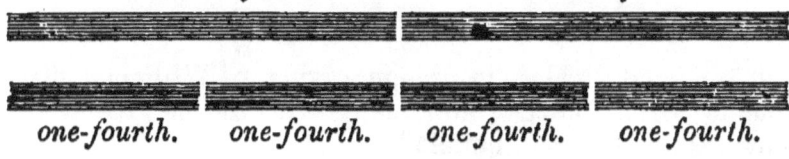

one-fourth.   *one-fourth.*   *one-fourth.*   *one-fourth.*

The first of the lines below represents a yard divided into 3 equal parts; the second, a yard divided into 6 equal parts. From this it is seen that $\frac{1}{3}$ is equal to $\frac{2}{6}$; and that $\frac{2}{3}$ are equal to $\frac{4}{6}$.

one-third.   *one-third.*   *one-third.*

1-*sixth.*   1-*sixth.*   1-*sixth.*   1-*sixth.*   1-*sixth.*   1-*sixth.*

REMARK.—The Numerator and Denominator are called the *terms* of the fraction; when these are the smallest numbers of which the fraction admits, it is said to be in its *lowest terms.*

1. Reduce $\frac{3}{6}$ to its lowest terms. *Divide both terms by the greatest number (3), that will exactly divide both.*

2. Reduce $\frac{4}{6}$ to its lowest terms. $\frac{2}{8}$ to its lowest terms. $\frac{6}{8}$ to its lowest terms.

3. Reduce to their lowest terms, $\frac{3}{9}$, $\frac{6}{9}$.

4. Reduce to their lowest terms, $\frac{2}{10}$, $\frac{5}{10}$.

5. Reduce to their lowest terms, $\frac{4}{10}$, $\frac{6}{10}$, $\frac{8}{10}$, $\frac{9}{10}$.

6. Reduce to their lowest terms, $\frac{3}{12}$, $\frac{4}{12}$, $\frac{6}{12}$, $\frac{8}{12}$, $\frac{9}{12}$, $\frac{10}{12}$.

7. Reduce to their lowest terms, $\frac{7}{14}$, $\frac{5}{15}$, $\frac{6}{15}$, $\frac{9}{15}$, $\frac{10}{15}$, $\frac{12}{15}$.

8. Reduce to their lowest terms, $\frac{4}{18}, \frac{6}{18}, \frac{8}{18}, \frac{10}{18}, \frac{12}{18}, \frac{14}{18}$.
9. Reduce to their lowest terms, $\frac{2}{18}, \frac{4}{18}, \frac{6}{18}, \frac{8}{18}, \frac{10}{18}, \frac{12}{18}$.
10. Reduce to their lowest terms, $\frac{5}{20}, \frac{3}{21}, \frac{6}{24}, \frac{10}{25}, \frac{10}{30}, \frac{15}{35}$.
11. Reduce to their lowest terms, $\frac{14}{42}, \frac{16}{56}, \frac{18}{63}, \frac{24}{80}, \frac{60}{72}, \frac{55}{88}$.

SECTION XVII.

LESSON I.

1. David divided an orange, giving to William 1-fifth, and to Sarah 2-fifths: how many fifths did he give away? how many fifths had he left?

2. After taking from any thing 3-fifths of itself, what part will be left?

3. John bought a quart of chestnuts: he gave 2-sixths to Mary, and 3-sixths to Eliza: how many sixths did he give away? what part had he left?

4. One-fifth, 2-fifths, and 7-fifths of an orange, are how many fifths? how many oranges?

5. How much are $\frac{1}{5}, \frac{3}{5}$, and $\frac{7}{5}$?

6. One-eighth, 3-eighths, and 7-eighths of a dollar are how many eighths? how many dollars?

7. What is the sum of $\frac{2}{9}, \frac{4}{9}$, and $\frac{7}{9}$?

8. Daniel's mother gave him $\$\frac{1}{2}$, and his uncle $\$\frac{1}{3}$: how many sixths of a dollar did he receive from both?

ANALYSIS.—*One-half is 3-sixths, and 1-third is 2-sixths: 3-sixths and 2-sixths are 5-sixths. Ans. 5-sixths of a dollar.*

The fractions are reduced to a common denominator before adding, because you can not add things of different kinds. This is the first step in adding or subtracting fractions.

9. Lucy divided an orange, giving to her sister 1-third, and to her brother 1-sixth: how many sixths did she give? how many sixths did she have left?

FRACTIONS.

10. James bought a lemon, and gave to Lucy 1-eighth, and to Susan 1-fourth: how many eighths did he give? what part did he retain?

11. William cut a pine-apple, and gave to Mary 1-third, and to Eliza 1-ninth: how many ninths did he give to both? what part did he retain?

12. Thomas bought a copy-book for $\$\frac{1}{10}$, and a reader for $\$\frac{1}{2}$: how many tenths of a dollar did they both cost?

13. David gave 1-fourth of a melon to Eliza, 1-third to his mother, and kept the remainder: how many twelfths did he give? what part did he retain?

14. I bought $1\frac{1}{4}$ yards at one store, and $2\frac{1}{3}$ yards at another: how many yards did I purchase?

15. I planted $2\frac{1}{2}$ acres of ground in corn; $8\frac{2}{5}$ acres in oats: how many acres were in both pieces?

16. John bought a knife for $\$\frac{1}{2}$, a slate for $\$\frac{1}{8}$, and a book for $\$\frac{5}{8}$: how much did the whole cost?

17. Add $\frac{1}{2}$ and $\frac{3}{5}$. $\frac{1}{2}$ and $\frac{3}{8}$. $\frac{2}{3}$ and $\frac{3}{4}$.
18. Add $\frac{1}{3}$ and $\frac{2}{5}$. $\frac{1}{4}$ and $\frac{1}{5}$. $\frac{1}{4}$ and $\frac{1}{5}$.
19. Add $\frac{1}{2}$ and $\frac{1}{7}$. $\frac{3}{4}$ and $\frac{5}{7}$. $\frac{3}{4}$ and $\frac{5}{8}$.
20. Add $\frac{5}{6}$ and $\frac{4}{5}$. $\frac{5}{6}$ and $\frac{5}{8}$. $\frac{5}{7}$ and $\frac{5}{9}$.
21. Add $1\frac{3}{4}$ and $2\frac{1}{3}$. $\frac{1}{2}$, $\frac{1}{3}$, $\frac{1}{4}$, and $\frac{1}{5}$.
22. Add $3\frac{2}{3}$ and $4\frac{5}{6}$. $\frac{2}{3}$, $\frac{1}{4}$, $\frac{1}{5}$, and $\frac{1}{2}$.
23. Add $4\frac{3}{7}$ and $5\frac{1}{3}$. $1\frac{1}{2}$, $2\frac{3}{4}$, $\frac{4}{5}$, and $6\frac{2}{10}$.
24. Add $5\frac{3}{8}$ and $4\frac{4}{9}$. $\frac{5}{8}$, $3\frac{1}{3}$, 4, and $5\frac{5}{8}$.

LESSON II.

1. Mary had 3-fourths of an orange; she gave her sister 1-fourth: how many fourths had she left?

2. If you take 1-fifth from 3-fifths, what will be left?

3. I bought 5-sixths of a quart of nuts, and gave 3-sixths to my sister: how many sixths had I left?

4. One is how many sixths? If you take 5-sixths from 1, what will be left?

5. Two is how many fifths? If you take 3-fifths from 2, what will be left?

6. If you take $\frac{1}{8}$ from $\frac{3}{8}$, what will be left?
7. $\frac{3}{7}$ from $\frac{5}{7}$?
8. $\frac{3}{8}$ from $\frac{7}{8}$?
9. $\frac{2}{9}$ from $\frac{7}{9}$?
10. $\frac{3}{10}$ from $\frac{9}{10}$?
11. $\frac{3}{4}$ from 1?
12. $\frac{5}{8}$ from 2?

13. Thomas divided an orange, giving his brother 1-half, and his sister 1-third: how many sixths did each receive? how much more the brother than the sister?

14. If a bushel of wheat cost $\$\frac{1}{2}$, and of corn $\$\frac{1}{4}$, how much will the wheat cost more than the corn?

15. If you take $\frac{1}{2}$ from $\frac{3}{4}$, what will remain?

16. Joseph bought a quart of chestnuts, and gave 1-half of them to his mother, and 1-sixth to his sister: how many sixths did he give his mother more than his sister?

17. Take $\frac{1}{2}$ and $\frac{1}{6}$ from 1, what will be left?

18. Jane divided an orange, giving to her sister 3-eighths, and to her brother 1-fourth: to which did she give the most? what part of the orange was left?

19. A man having 72 miles to travel, went 1-third the distance the first day, 2-ninths the second, the remainder the third day: how many ninths of the distance did he travel further on the first day, than on the second? what part did he travel the last day?

20. If you take $\frac{2}{9}$ from $\frac{1}{3}$, what will be left? If you take $\frac{2}{9}$ and $\frac{1}{3}$ from 1, what will be left?

21. If you take $\frac{1}{5}$ from $\frac{1}{2}$, what will be left?

ANALYSIS.—1-*fifth* is 2-*tenths*, and 1-*half* is 5-*tenths*; 2-*tenths* from 5-*tenths* leave 3-*tenths*. Ans. 3-*tenths*.

What will be left if you take

22. $\frac{1}{4}$ from $\frac{1}{3}$? $\frac{1}{5}$ from $\frac{1}{3}$? $\frac{1}{5}$ from $\frac{1}{4}$?
23. $\frac{1}{6}$ from $\frac{1}{2}$? $\frac{2}{3}$ from $\frac{1}{2}$? $\frac{1}{2}$ from $\frac{3}{8}$?
24. $\frac{1}{3}$ from $\frac{3}{4}$? $\frac{1}{4}$ from $\frac{3}{5}$? $\frac{3}{8}$ from $\frac{5}{8}$?

25. $\frac{3}{7}$ from $\frac{2}{3}$? $\frac{1}{2}$ from $\frac{7}{9}$? $\frac{1}{4}$ from $\frac{3}{5}$?
26. $\frac{3}{6}$ from $\frac{5}{7}$? $\frac{5}{7}$ from $\frac{3}{4}$? $\frac{3}{5}$ from $\frac{5}{6}$?
27. $\frac{1}{3}$ from $\frac{3}{8}$? $\frac{5}{6}$ from $\frac{6}{7}$? $\frac{1}{3}$ from $2\frac{1}{2}$?

LESSON III.

1. Mary divided a quart of pecans, giving Ann 1-third, and Jane 1-fourth of them: what part had she left?

2. A farmer had $1\frac{1}{2}$ bushels of wheat: he gave to 1 poor man $\frac{1}{2}$ of a bushel, and to another $\frac{1}{3}$ of a bushel: how much wheat was left?

3. James had $\frac{7}{8}$ of a pound of raisins: he gave to his brother $\frac{1}{2}$ of a pound, and to his sister $\frac{1}{4}$ of a pound: how much had he left?

4. If from $3\frac{1}{2}$ bushels of corn, $1\frac{1}{2}$ bushels be taken, how much will there be left?

5. A lady bought $3\frac{1}{3}$ yards of muslin at one store, and $2\frac{1}{4}$ yards at another: after using $1\frac{1}{2}$ yards, how much had she left?

6. William's father gave him $\$\frac{5}{8}$: he gave to a poor person $\$\frac{1}{8}$, for apples $\$\frac{1}{16}$, and for a book $\$\frac{1}{4}$: what part of a dollar had he left?

7. James's mother gave him a book: he read the first day $\frac{1}{6}$, the second $\frac{1}{4}$, the third $\frac{1}{2}$, and the fourth the remainder: what part did he read the fourth day?

8. A farmer has a flock of 84 sheep in 4 fields: the first contains $\frac{1}{6}$, the second $\frac{1}{2}$, and the third $\frac{1}{4}$ of them: what part does the fourth field contain?

9. Daniel spends $\frac{1}{3}$ of his time in sleep, $\frac{1}{4}$ of it at school, $\frac{1}{12}$ in reading, and $\frac{1}{24}$ in learning music: what part of his time is not employed?

10. A pole is standing in a pond; $\frac{1}{2}$ of it is in the air, and $\frac{1}{3}$ in the water: what part is in the earth?

11. A student devotes $\frac{1}{4}$ of his time to sleep, $\frac{1}{3}$ to study, $\frac{1}{24}$ to reading, $\frac{1}{8}$ to exercise, and $\frac{1}{12}$ to deeds of charity: what part of his time is unemployed?

12. One-third of an orchard is apple trees, ¼ pear trees, ⅛ plum trees, 1/12 quince trees, and the remainder peach trees: if the orchard contain 96 trees, how many are there of each kind?

13. After spending ½ and ⅓ of my money, and losing 1/12, I had $8 remaining: how much had I at first?

14°. The difference between ⅝ and ½ of my money is $3: how much have I?

15°. I ate 5/7 of my peaches, gave away 10, and have 10 left: how many had I at first?

SECTION XVIII.

LESSON I.

1. A father gave each of his two sons half a dollar: how much did he give them both?

2. A mother gave each of her 3 children half an orange: how many half oranges did it take? how many oranges? why?

3. What are 3 times 1-half? 4 times 1-half?

4. John fed 5 horses, giving to each half a peck of oats: how many half pecks did it take? how many pecks? why?

5. What are 5 times 1-half? 6 times 1-half?

6. What are 8 times 1-half? 9 times 1-half?

7. James gave 1-third of an orange to each of his sisters: how much did he give to both? why?

8. If a man can eat 1-third of a pound of meat in 1 day, how much can he eat in 3 days? why?

9. What are 4 times 1-third? 5-times 1-third?

10. John gave 2-thirds of a pine-apple to each of his 2 brothers: how many thirds did he give to both? How many pine-apples did it take? *Ans.* 1 *and* 1-*third*.

FRACTIONS.

11. What are 4 times 2-thirds? 5 times 2-thirds?

12. Thomas gave 1-fourth of an apple to each of his 3 playmates: how many fourths did it take? why?

13. If 1 bushel of oats cost 1-fourth of a dollar, how much will 4 bushels cost? why?

14. Charles gave 3-fourths of a pint of chestnuts to each of his 2 brothers: how many fourths of a pint did it take? how many pints? why?

15. Mary gave 3-fourths of an orange to each of her three brothers: how many fourths of an orange did it take? how many oranges? why?

16. What are 5 times 3-fourths?

ANALYSIS.—5 *times* 3-*fourths are* 15-*fourths*; 15-*fourths are as many ones as* 4-*fourths are contained times in* 15-*fourths*; 4-*fourths in* 15-*fourths*, 3 *and* 3-*fourths times.* Ans. 3 and 3-fourths.

17. What are 6 times $\frac{3}{4}$? 7 times $\frac{3}{4}$? 8 times $\frac{3}{4}$?
18. What are 3 times $\frac{1}{5}$? 5 times $\frac{2}{5}$? 6 times $\frac{3}{5}$?
19. What are 3 times $\frac{1}{7}$? 4 times $\frac{2}{7}$? 5 times $\frac{4}{7}$?
20. What are 5 times $\frac{1}{8}$? 2 times $\frac{3}{8}$? 4 times $\frac{7}{8}$?
21. What are 2 times $\frac{1}{9}$? 4 times $\frac{2}{9}$? 5 times $\frac{4}{9}$?
22. What are 8 times $\frac{3}{9}$? 6 times $\frac{6}{9}$? 7 times $\frac{8}{9}$?
23. What are 5 times $\frac{7}{6}$? 8 times $\frac{3}{5}$? 9 times $\frac{2}{3}$?
24. What are 7 times $\frac{5}{8}$? 6 times $\frac{7}{7}$? 8 times $\frac{3}{2}$?
25. What are 3 times $\frac{5}{4}$? 4 times $\frac{6}{5}$? 6 times $\frac{8}{7}$?
26. What are 7 times $\frac{5}{8}$? 8 times $\frac{5}{2}$? 9 times $\frac{4}{3}$?
27. What are 3 times $\frac{3}{10}$? 4 times $\frac{5}{10}$? 7 times $\frac{6}{10}$?

LESSON II.

1. William gave an orange and a half to each of his 2 sisters: how many oranges did it take?

2. What are 2 times 1 and 1-half?

3. What are 3 times $1\frac{1}{2}$? 4 times $1\frac{1}{2}$? 5 times $1\frac{1}{2}$?

ANAL.—*3 times 1 are 3, and 3 times 1-half are 3-halves, equal 1 and 1-half; this added to 3 makes 4 and 1-half.*

4. If 1 bushel of wheat cost $1 and 1-third of a dollar, what will 2 bushels cost?

5. How many are 3 times $1\tfrac{1}{3}$? 2 times $2\tfrac{2}{3}$?
6. How many are 3 times $3\tfrac{1}{3}$? 4 times $4\tfrac{1}{3}$?
7. How many are 5 times $2\tfrac{2}{3}$? 6 times $3\tfrac{2}{3}$?
8. How many are 8 times $3\tfrac{1}{3}$? 9 times $4\tfrac{2}{3}$?

9. If 1 bushel of barley cost $1 and 1-fourth of a dollar, what will 3 bushels cost? 4 bushels?

10. How many are 5 times $1\tfrac{1}{4}$? 6 times $1\tfrac{1}{4}$?
11. How many are 2 times $1\tfrac{3}{4}$? 3 times $2\tfrac{1}{4}$?
12. How many are 4 times $3\tfrac{1}{4}$? 5 times $3\tfrac{3}{4}$?
13. How many are 6 times $3\tfrac{1}{4}$? 8 times $3\tfrac{3}{4}$?
14. How many are 7 times $2\tfrac{1}{4}$? 9 times $2\tfrac{3}{4}$?
15. How many are 10 times $1\tfrac{3}{4}$? 10 times $3\tfrac{1}{4}$?
16. How many are 11 times $2\tfrac{1}{4}$? 12 times $3\tfrac{3}{4}$?

17. If a family consume 3 and 1-fifth barrels of flour in one month, how much will they require for 3 months?

18. How many are 4 times $3\tfrac{2}{5}$? 5 times $3\tfrac{3}{5}$?
19. How many are 2 times $6\tfrac{3}{5}$? 3 times $2\tfrac{4}{5}$?
20. How many are 6 times $4\tfrac{1}{5}$? 6 times $3\tfrac{1}{5}$?
21. How many are 7 times $4\tfrac{2}{5}$? 8 times $3\tfrac{2}{5}$?
22. How many are 9 times $1\tfrac{4}{5}$? 9 times $3\tfrac{1}{5}$?

SECTION XIX.

LESSON I.

1. If you divide an apple into two equal parts, what is 1 part called? what is 1-half of 1?

FRACTIONS.

2. If you divide 3 apples between 2 boys, how many apples will each have? how will you divide them? what is 1-half of 3?

ANALYSIS.—1-*half of* 1 *is* 1-*half, and* 1-*half of* 3 *is* 3 *times* 1-*half of* 1, *which are* 3-*halves;* 3-*halves are as many ones as* 2-*halves are contained times in* 3-*halves;* 2-*halves in* 3-*halves,* 1 *and* 1-*half times. Ans.* 1 *and* 1-*half.*

3. John bought 5 oranges, and divided them between his 2 sisters: how many oranges did each have?

4. What is 1-half of 6? ½ of 7? ½ of 8? ½ of 9? of 11? of 13?

5. If cloth is $1 a yard, what must be paid for 1-third of a yard? what is 1-third of 1?

6. James has 2 oranges to divide among his 3 sisters: how must he divide them, and what part of an orange will each have? what is 1-third of 2?

7. If 3 bushels of pears cost $4, how much is that a bushel? what is 1-third of 4?

ANALYSIS.—1-*third of* 4 *is the same as* 4 *times* 1-*third of* 1, *which are* 4-*thirds;* 4-*thirds are as many ones as* 3-*thirds are contained times in* 4-*thirds:* 3-*thirds in* 4-*thirds* 1 *and* 1-*third times. Ans.* 1 *and* 1-*third.*

8. A carpenter receives $5 for 3 days' work: how much is that a day? what is 1-third of 5?

9. What is 1-third of 6? ⅓ of 7? ⅓ of 8? ⅓ of 9? of 10? of 11? of 13?

10. What is 1-third of 16? of 17? of 18? of 19? of 20? of 21? of 22?

11. What is 1-third of 23? of 24? of 25? of 26? of 27? of 28? of 31?

12. If 4 bushels of oats cost $1, what part of a dollar will 1 bushel cost? what is 1-fourth of 1?

13. A mother divided 3 pears equally among her 4 children, what part of a pear did each receive?

14. What is 1-fourth of 4? ¼ of 5? ¼ of 6? of 7? of 8? of 9? of 10? of 11?

15. What is 1-fourth of 12? of 13? of 15? of 17? of 19? of 23?

16. What is 1-fourth of 25? of 26? of 27? of 28? of 29? of 30? of 31?

17. If a melon cost 5 cents, what part of a melon can you buy for 1 cent? what is 1-fifth of 1?

18. James had 2 pine-apples, and divided them among 5 of his companions: what part of a pine-apple did each have? what is 1-fifth of 2?

19. What is 1-fifth of 3? $\frac{1}{5}$ of 4? $\frac{1}{5}$ of 5? of 6? of 7? of 8.? of 9? of 10? of 11? of 12?

20. What is 1-sixth of 1? $\frac{1}{6}$ of 2? $\frac{1}{6}$ of 3? of 4? of 5? of 6? of 7? of 8? of 10? of 15?

21. What is 1-ninth of 2? $\frac{1}{9}$ of 4? $\frac{1}{9}$ of 7? of 8? of 9? of 12? of 15? of 17? of 19?

22. How do you find one-half of any thing? one-third? one-fourth? one-fifth? one-eighth?

LESSON II.

1. If 3 bushels of wheat cost $1, what part of $1 will 1 bushel cost? what part will 2 bushels cost?

2. What is 1-third of 1? what are 2-thirds of 1?

3. If 3 bushels of wheat cost $2, what part of $1 will 1 bushel cost? what part will 2 bushels cost?

ANAL.—*1 bushel will cost 1-third of $2, which is 2-thirds; and 2 bushels will cost twice as much as 1 bushel; that is, 2 times 2-thirds, which are 4-thirds of $1, equal $1 and 1-third of a dollar. Ans. $1 and 1-third.*

4. If 3 barrels of cider cost $4, what part of a dollar will 1 barrel cost? what part will 2 barrels cost?

5. What is 1-third of 4?   2-thirds of 4?
6. What is 1-third of 5?   2-thirds of 5?

FRACTIONS. 103

7. What is 1-third of 6?   2-thirds of 6?
8. What is 1-third of 7?   2-thirds of 7?
9. What is 1-third of 8?   2-thirds of 8?
10. What is 1-third of 9?   2-thirds of 9?
11. What is 1-third of 10?   2-thirds of 10?

12. If 4 barrels of apples cost $3, what part of a dollar will 1 barrel cost? 2 barrels? 3 barrels?

13. If 5 apples cost 2 cents, what part of a cent will 1 apple cost? 2 apples? 3 apples? 4 apples?

14. What are 2-fifths of 13?   3-fifths of 13?
15. What is 1-sixth of 7?   5-sixths of 7?
16. What are 3-sixths of 10?   4-sixths of 10?
17. What is 1-seventh of 9?   3-sevenths of 9?
18. What are 2-sevenths of 11?   3-sevenths of 12?
19. What is 1-eighth of 13?   3-eighths of 13?
20. What are 5-thirds of 4?   7-thirds of 5?
21. What are 7-eighths of 10?   9-eighths of 3?
22. What are 3-tenths of 7?   7-tenths of 3?
23. What are 5-ninths of 11?   9-fifths of 7?

24. Which is the greater $\frac{1}{3}$ of 2, or $\frac{2}{3}$ of 1? $\frac{1}{4}$ of 3, or $\frac{3}{4}$ of 1?

LESSON III.

1. James bought 3 lemons for 7 cents: how much was that apiece?

2. William bought 5 quarts of chestnuts for 18 cents: at that rate, what was the cost of 2 quarts?

3. If 7 pounds of cheese sell for 40 cents, how much should 5 pounds sell for?

4. Bought 6 yards of muslin; gave my mother 3-fifths, and kept the remainder: how many yards had each?

5. If a quantity of provisions serve 2 men 7 days, how long will it last 1 man? how long 3 men?

6. If 4 men perform a job of work in 8 days, how long will it require 5 men?

ANALYSIS.—*It will take 1 man 4 times as long as 4 men; and 5 men 1-fifth as long as 1 man.*

7. If a barrel of cider will last 5 men 8 days, how long will it last 3 men?

8. If 1 barrel of flour serve 8 persons 10 days, how long will it last 11 persons?

9. If 7 men can do a piece of work in 5 days, how long will it require 8 men?

10. If 2 men build a wall in 12 days, how long will it take 7 men?

11. If it requires 11 days, of 8 hours each, to do a job, how many will be required of 10 hours each?

12. A man paid 37 cents for riding 8 miles: at the same rate, what will it cost to ride 11 miles?

13. Two pipes of a certain size will empty a cistern in 17 minutes: in what time will 3 pipes empty it?

14. If 18 bushels of oats last 5 horses 1 week, how many bushels will 7 horses require?

15. If a laborer receive 5 bushels of wheat for 7 days' work, how much should he receive for 11 days?

16. If a carpenter earn $8 in 5 days, how much will he earn in 9 days?

17. If 3 yards of cloth cost $13, what cost 8 yards?

18. A pole, 18 feet long, is 2-sevenths in the earth, the rest in the air: what is the length of each part?

19. Three men, A, B and C, found a bag containing $15: A got 2-ninths, B 1-third, and C the remainder: what was the share of each?

20. If $\frac{1}{2}$ a yard of cloth cost $4, what cost 6 yards?

21. If 3 ounces cost 36 cents, what should be charged for $\frac{2}{3}$ of an ounce?

22. If $\frac{2}{3}$ of a number equal 18, what is the $\frac{4}{5}$ of it?

23. How much must a man earn a day to receive $72 for 8 weeks, 6 days to a week?

SECTION XX.

LESSON I.

ILLUSTRATION.—If 1-half of a yard of tape be divided into 2 equal parts, one of the parts is 1-half of 1-half; that is, 1-fourth of the whole yard.

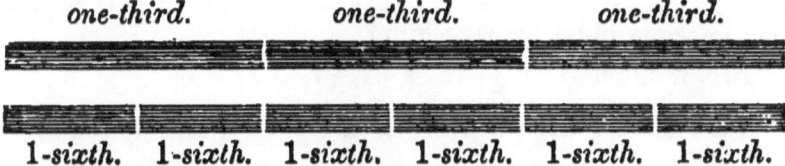

If 1-third of a yard be divided into two equal parts, one of the parts is 1-sixth of the whole yard; that is, 1-half of 1-third, is 1-sixth.

1. Mary having 1-half of an orange, gave her brother 1-half of what she had: what part of a whole orange did she give him?

2. James divided 1-third of an apple equally between his 2 brothers; what part did each receive?

3. What is $\frac{1}{2}$ of $\frac{1}{2}$? $\frac{1}{2}$ of $\frac{1}{3}$?

ANALYSIS.—*1-half of 1-third is the fraction obtained by dividing a unit into 3 equal parts, and one of these parts into 2 equal parts; but if a unit* (one) *be thus divided, there will be 6 parts, and each part will be 1-sixth of the whole; therefore, 1-half of 1-third is 1-sixth.*

4. Thomas divided 1-half of a lemon equally between his 3 sisters: what part did each receive?

5. If 1-fourth of an orange be divided into 2 equal parts, what is 1 of the parts called?

6. What is $\frac{1}{2}$ of $\frac{1}{4}$? $\frac{1}{4}$ of $\frac{1}{2}$?

7. If 1-third of an apple be cut into 3 equal parts, what part of the apple will each piece be?

8. If each half of an apple be divided into 5 equal parts, how many parts will there be? what is 1 part called? what is $\frac{1}{5}$ of $\frac{1}{2}$?

9. If you divide an orange into 4 equal parts, and cut each part into 3 pieces, what is 1 piece called?

What single Fraction equals

10. $\frac{1}{3}$ of $\frac{1}{4}$? $\frac{1}{4}$ of $\frac{1}{3}$? $\frac{1}{2}$ of $\frac{1}{7}$? $\frac{1}{3}$ of $\frac{1}{5}$?
11. $\frac{1}{4}$ of $\frac{1}{4}$? $\frac{1}{3}$ of $\frac{1}{6}$? $\frac{1}{4}$ of $\frac{1}{5}$? $\frac{1}{3}$ of $\frac{1}{7}$?
12. $\frac{1}{4}$ of $\frac{1}{6}$? $\frac{1}{3}$ of $\frac{1}{8}$? $\frac{1}{5}$ of $\frac{1}{5}$? $\frac{1}{3}$ of $\frac{1}{9}$?
13. $\frac{1}{4}$ of $\frac{1}{7}$? $\frac{1}{5}$ of $\frac{1}{6}$? $\frac{1}{4}$ of $\frac{1}{8}$? $\frac{1}{5}$ of $\frac{1}{7}$?

14. Thomas has 2-thirds of an apple, and wishes to give his brother 1-half of what he has: what part of the whole apple must he give him?

15. What is $\frac{1}{2}$ of $\frac{1}{4}$? $\frac{1}{2}$ of $\frac{2}{3}$?

ANALYSIS.—1-*half of* 2-*thirds is* 2 *times* 1-*half of* 1-*third*; 1-*half of* 1-*third is* 1-*sixth*; 2 *times* 1-*sixth are* 2-*sixths*.

16. Daniel has 3-fifths of a melon to divide equally between his brother and sister: how must he divide it, and what part of the whole will each receive?

17. What is $\frac{1}{2}$ of $\frac{2}{3}$? $\frac{1}{4}$ of $\frac{2}{3}$? $\frac{1}{3}$ of $\frac{5}{6}$?
18. What is $\frac{1}{4}$ of $\frac{3}{4}$? $\frac{1}{5}$ of $\frac{5}{6}$? $\frac{1}{6}$ of $\frac{3}{4}$?
19. What is $\frac{1}{6}$ of $\frac{4}{7}$? $\frac{1}{7}$ of $\frac{5}{8}$? $\frac{1}{9}$ of $\frac{5}{7}$?
20. What is $\frac{1}{7}$ of $\frac{4}{5}$? $\frac{1}{8}$ of $\frac{4}{9}$? $\frac{1}{9}$ of $\frac{5}{9}$?

21. Edward has 4-fifths of a melon, and wishes to give his sister 2-thirds of what he has: what part of the whole melon will she receive?

FRACTIONS. 107

22. What is $\frac{1}{2}$ of $\frac{9}{10}$? $\frac{1}{3}$ of $\frac{6}{8}$? $\frac{2}{3}$ of $\frac{4}{5}$?

ANALYSIS.—2-*thirds of* 4-*fifths are* 2 *times* 1-*third of* 4-*fifths, and* 1-*third of* 4-*fifths is* 4 *times* 1-*third of* 1-*fifth.*

23. What are $\frac{3}{4}$ of $\frac{2}{3}$? $\frac{2}{3}$ of $\frac{3}{5}$? $\frac{3}{4}$ of $\frac{6}{5}$?
24. What are $\frac{2}{3}$ of $\frac{5}{6}$? $\frac{3}{4}$ of $\frac{2}{7}$? $\frac{3}{5}$ of $\frac{3}{8}$?
25. What are $\frac{2}{5}$ of $\frac{3}{7}$? $\frac{5}{8}$ of $\frac{4}{7}$? $\frac{2}{8}$ of $\frac{7}{10}$?
26. What are $\frac{3}{5}$ of $\frac{8}{9}$? $\frac{2}{7}$ of $\frac{5}{11}$? $\frac{3}{8}$ of $\frac{5}{7}$?

27. If 1 yard of cloth is worth $2\frac{1}{2}$ bushels of wheat, what is 1-half of a yard worth?

ANALYSIS.—2 *and* 1-*half are* 5-*halves, and* 1-*half of* 5-*halves is* 5 *times* 1-*half of* 1-*half, or* 5-*fourths, equal* 1 *and* 1-*fourth. Ans.* 1 *and* 1-*fourth bushels.*

What single Fraction will represent

28. $\frac{1}{3}$ of $2\frac{1}{2}$? $\frac{1}{2}$ of $1\frac{1}{4}$? $\frac{1}{3}$ of $1\frac{3}{4}$? $\frac{1}{4}$ of $2\frac{1}{5}$?
29. $\frac{1}{4}$ of $2\frac{1}{4}$? $\frac{1}{5}$ of $3\frac{1}{4}$? $\frac{1}{6}$ of $4\frac{2}{3}$? $\frac{1}{7}$ of $5\frac{1}{6}$?
30. $\frac{2}{3}$ of $1\frac{1}{2}$? $\frac{3}{4}$ of $1\frac{2}{3}$? $\frac{2}{7}$ of $1\frac{1}{3}$? $\frac{3}{7}$ of $1\frac{1}{4}$?
31. $\frac{3}{4}$ of $2\frac{1}{5}$? $\frac{2}{5}$ of $4\frac{1}{3}$? $\frac{3}{8}$ of $2\frac{2}{7}$? $\frac{5}{8}$ of $3\frac{1}{2}$?

LESSON II.

1. A person owning $\frac{1}{3}$ of a ship, sold $\frac{1}{2}$ of his share: what part of the whole ship did he sell?

2. If 3 yards of cloth cost $\$2\frac{1}{3}$, what cost 2 yards?

ANALYSIS.—1 *yard will cost* 1-*third as much as* 3 *yards, and* 2 *yards will cost* 2 *times as much as* 1 *yard; that is,* 2-*thirds as much as* 3 *yards.*

3. If 2 yards of cloth cost $\$1\frac{1}{4}$, what cost 3 yards?

4. If 3 yards of cloth cost $\$5\frac{1}{2}$, what part of that sum will 2 yards cost?

5. If 5 gallons of molasses cost $\$3\frac{2}{3}$, what part of that sum will 3 gallons cost?

6. If 7 pounds of sugar cost 1\frac{1}{7}$, what cost 4 pounds?

7. If 8 pounds of butter cost 1\frac{1}{4}$, what cost 7 pounds?

8. If 7 yards of cloth cost 5\frac{2}{3}$, what will be the cost of 3 yards? of 4 yards?

9. If 3 barrels of cider cost 4\frac{2}{3}$, what part of that sum will 5 barrels cost?

10. If 5 gallons of oil cost 2\frac{2}{5}$, what part of that sum will 7 gallons cost?

11. If 2 gallons of molasses cost 1\frac{2}{5}$, what will be the cost of 3 gallons?

12. If 3 bottles of wine cost 2\frac{1}{3}$, what will be the cost of 8 bottles? of 10 bottles?

13. If 2 men do a job of work in 3$\frac{1}{2}$ days, how many days will it take 1 man? 3 men?

14. A man can do a job in 3$\frac{2}{3}$ days, of 10 hours each: how many days will it take of 7 hours each?

15. If a man can do a piece of work in 15$\frac{3}{4}$ days, working 5 hours a day, how many days will it take, working 8 hours a day?

SECTION XXI.

LESSON I.

1. A boy had 2-thirds of an orange, which he divided equally between his two sisters: what part of an orange did each receive?

ANALYSIS.—*Since 2 apples divided by 2 give 1 apple, and $2 divided by 2 give $1, it follows that 2-thirds divided by 2 must give 1-third. Each, therefore, must have received 1-third of an orange.*

2. How often is 3 contained in 3-fourths?

3. If 2 and 2-thirds dollars be divided among 4 men, what is each man's share?

FRACTIONS.

4. If $2\tfrac{4}{5}$ be divided by 7, what will be the result?
5. How many times 6, in 3 and 3-fifths?
6. How many thirds of 9, are contained in $3\tfrac{6}{8}$?
7. Divide 6 and 3-fourths cents among 9 boys.
8. A man having $10\tfrac{5}{7}$ acres of land, divided it equally between his 5 children: how much did each receive?
9. If $4\tfrac{5}{7}$ be divided by 11, what will be the quotient?
10. If $\$\tfrac{2}{5}$ be equally divided among 3 men, what part of a dollar will each get?
11. How divide a fraction by any whole number?

LESSON II.

1. If 1 apple cost $\tfrac{1}{2}$ a cent, how many will 1 cent buy? why? *Ans. Because there are 2 half-cents in 1 cent.*
2. If 1 pear cost 1-third of a cent, how many pears can you buy for 1 cent? why?
3. When 1 bushel of wheat costs $\$\tfrac{1}{2}$, how many bushels can you buy for $\$1\tfrac{1}{2}$?
4. How often is $\tfrac{1}{2}$ contained in $2\tfrac{1}{2}$ or $\tfrac{5}{2}$?

ANALYSIS.—*One-half is contained in* FIVE-*halves, as often as* 1 *is contained times in* 5; *that is,* 5 *times: for, if* 1 *apple is contained in* 5 *apples* 5 *times;* 1 *cent in* 5 *cents,* 5 *times;* 1-*half must be contained in* 5-*halves,* 5 *times.*

5. If 1 lemon cost half a cent, how many lemons can be bought for 5 cents?
6. If 1 apple cost 1-third of a cent, how many apples can John buy with 4 and 1-third cents?
7. If 1 peach cost 2-thirds of a cent, how many peaches can you purchase for 4 and 1-third cents?

ANALYSIS.—*4 and* 1-*third are* 13-*thirds; and* TWO-*thirds are contained in* THIRTEEN-*thirds as often as* 2 *is contained in* 13, *that is,* 6 *and* 1-*half times. Ans.* $6\tfrac{1}{2}$ *peaches.*

8. I distributed $2\frac{2}{3}$ bushels of wheat among a number of poor persons, giving to each, 2-thirds of a bushel: how many persons were there?

9. At $\$\frac{1}{4}$ a yard, how many yards of calico can be purchased for $\$3\frac{3}{4}$?

10. At $\$\frac{3}{4}$ a yard, how many yards of cloth can be purchased for $\$3\frac{1}{4}$?

11. If a lemon cost 3-fourths of a cent, how many can be purchased for $3\frac{3}{4}$ cents? for $5\frac{1}{4}$ cents?

12. One bushel of rye is worth 3-fourths of a bushel of wheat: how many bushels of rye can be bought with $4\frac{3}{4}$ bushels of wheat? with $8\frac{3}{4}$ bushels?

13. At 1-fifth of a cent each, how many cherries can Mary purchase with 3 cents? why?

14. At $\$\frac{2}{5}$ a gallon, how many gallons of vinegar can you buy for $\$2\frac{2}{5}$? for $\$4\frac{1}{5}$?

15. How often is $\frac{3}{5}$ contained in $2\frac{2}{5}$? in $4\frac{3}{5}$? in $6\frac{1}{5}$?
16. How often is $\frac{1}{6}$ contained in $3\frac{5}{6}$? in $5\frac{1}{6}$? in $4\frac{4}{6}$?
17. How often is $\frac{3}{7}$ contained in 1? in $3\frac{2}{7}$? in $4\frac{1}{7}$?
18. How often is $\frac{5}{8}$ contained in 3? in $4\frac{3}{8}$? in $6\frac{1}{8}$?
19. At $\$\frac{1}{3}$ a yard, how much cloth can be had for $\$\frac{1}{2}$?

ANALYSIS.—*As many yards as 1-third of a dollar, the price of 1 yard, is contained times in 1-half of a dollar; 1-half is equal to 3-sixths, and 1-third to 2-sixths; 2-sixths in 3-sixths, 1 and 1-half times. Ans. 1 and 1-half yards.*

20. At $\$\frac{1}{2}$ a yard, how much gingham can be purchased for $\$\frac{3}{4}$?

21. How often is $\frac{1}{2}$ contained in $\frac{3}{4}$?

22. If 1 yard of calico cost $\$\frac{1}{2}$, how much can be purchased for $\$\frac{2}{3}$?

23. How often is $\frac{1}{2}$ contained in $\frac{3}{5}$?

NOTE.—Reduce the fractions to a common denominator, then divide their numerators.

FRACTIONS.

24. How often is $\frac{1}{2}$ contained in $\frac{5}{6}$? in $\frac{7}{8}$? in $1\frac{1}{2}$?
25. How often is $\frac{1}{3}$ contained in $\frac{4}{5}$? in $\frac{5}{7}$? in $1\frac{2}{13}$?
26. How often is $\frac{2}{3}$ contained in $\frac{3}{5}$? in $\frac{5}{8}$? in $1\frac{10}{12}$?
27. How often is $\frac{3}{4}$ contained in $\frac{6}{7}$? in $\frac{7}{8}$? in $\frac{9}{10}$?
28. How often is $1\frac{1}{2}$ contained in $\frac{3}{4}$? in $\frac{4}{5}$? in $2\frac{3}{4}$?
29. How often is $2\frac{1}{4}$ contained in $\frac{5}{6}$? in $\frac{5}{7}$? in $3\frac{7}{8}$?
30. How often is $3\frac{1}{5}$ contained in $\frac{3}{6}$? in $\frac{3}{7}$? in $5\frac{2}{8}$?

SECTION XXII.

LESSON I.

1. James gave his brother 1 apple and 1-half, which was 1-half of what he had: how many had he?

2. If 1-third of a yard cost 1\frac{1}{2}$, what cost a yard?

3. If a man travel 1 and 1-third miles in 1-fourth of an hour, how far can he travel in 1 hour?

4. Daniel bought 1-fifth of an orange for $3\frac{1}{4}$ cents: at that rate, what will a whole one cost?

5. If 1-fourth of a yard of tape cost 3 and 2-thirds cents, what will 1 yard cost?

ANALYSIS.—*3 and 2-thirds are 1-fourth of 4 times 3 and 2-thirds; 4 times 2-thirds are 8-thirds, equal 2 and 2-thirds; 4 times 3 are 12, which added to 2 and 2-thirds, make 14 and 2-thirds. Ans. 14 and 2-thirds.*

6. 5 and 1-third are 1-half of what number?
7. 7 and 3-fourths are 1-third of what number?
8. 5 and 3-eighths are 1-fifth of what number?
9. 3 and 4-sevenths are 1-seventh of what number?
10. 4 and 4-fifths are 1-sixth of what number?
11. 9 and 2-thirds are 1-eighth of what number?

12. If 2-thirds of a yard of ribbon cost 3 cents, what will 1-third cost? If 1-third of a yard cost 1 and 1-half cents, what will a yard cost?

13. One and 1-half is 1-third of what number? Three is 2-thirds of what number?

14. If 2-thirds of a barrel of flour cost $5, what cost 1-third of a barrel? If 1-third of a barrel cost $2½, what will a whole barrel cost?

15. Five is 2 times what number?

ANALYSIS.—5 *is* 2 *times* 1-*half of* 5; 1-*half of* 5 *is* 2 *and* 1-*half; therefore,* 5 *is* 2 *times* 2 *and* 1-*half.*

16. Five is 2-thirds of what number?

ANALYSIS.—5 *is* 2-*thirds of that number of which* 1-*half of* 5 *is* 1-*third*: 1-*half of* 5 *is* 2 *and* 1-*half, and* 2 *and* 1-*half are* 1-*third of* 3 *times* 2 *and* 1-*half, which are* 7 *and* 1-*half; therefore,* 5 *is* 2-*thirds of* 7 *and* 1-*half.*

17. If 3-fourths of a yard of velvet cost $4, what will 1-fourth of a yard cost? If 1-fourth cost $1⅓, what will a whole yard cost?

18. Four is 3-fourths of what number?

19. If 2-fifths of a quart of chestnuts cost 3 cents, what will 1-fifth cost? If 1-fifth of a quart cost 1 and 1-half cents, what will a whole quart cost?

20. Three is 2-fifths of what number?

21. If 4-fifths of a yard cost $7, what will 1-fifth cost? If 1-fifth of a yard cost $1¾, what cost a yard?

22. Seven is 4-fifths of what number?

23. If 3-eighths of a melon cost 4 cents, what will 1-eighth cost? what will a whole melon cost?

24. If 3-fifths of a pound of sugar cost 10 cents, what will 1-fifth cost? what will 1 pound cost?

25. If 5-sixths of a barrel of beef cost $6, what will 1-sixth cost? what will a barrel cost?

FRACTIONS. 113

26. Fifteen is 2-ninths of what number?
27. Fourteen is 3-eighths of what number?
28. Thirteen is 3-fourths of what number?
29. Seventeen is 8-ninths of what number?

30. If 1 yard and a half, that is, 3 half yards of ribbon cost 6 cents, what will a yard cost?

31. Six is 3-halves of what number?

ANALYSIS.—*Of that number of which 1-third of 6 is 1-half; 1-third of 6 is 2; 2 is 1-half of 2 times 2, which are 4; therefore, 6 is 3-halves of 4.*

32. If 1 yard and 1-third of a yard, that is, 4-thirds of a yard of cloth, cost $3, what will a yard cost?

33. If a man travel 9 miles in 1 hour and 2-sevenths of an hour, how far will he travel in 1 hour?

34. By selling a watch for $18, I gained 1-fourth of what it cost me: how much did it cost?

35. A grocer, by selling a lot of flour for $25, gained 1-fifth of what it cost him: what was the cost? how much did he gain?

36. If a man pay $6 for 1 and 1-third yards of cloth, what is the cost of 1 yard?

37. If a man receive $10 for 2 and 2-thirds days' work, how much is that per day?

38. If 2-thirds of a yard cost 4\frac{6}{7}$, what cost a yard?

39. Four and 6-sevenths, are 2-thirds of what?

ANALYSIS—*Of that number of which 1-half of 4 and 6-sevenths is 1-third.*

40. Four and 2-thirds are 2-fifths of what?
41. Three and 3-fourths are 5-sixths of what?
42. One and 3-fifths are 3-fifths of what?
43. Three and 2-thirds are 3-fourths of what?
44. Four and 4-fifths are 6-halves of what?

LESSON II.

1. A man gave to some poor persons $3, which was 2-fifths of his money: how much had he left?

2. A pole stands 5-sevenths in the air, and 5 feet in the ground: how long is the pole?

3. A man spent 2-fifths of his money, and had $10 left: how much did he spend?

4. At 3 yards for 2 cents, how many yards of tape can be bought for 7 cents?

5. At 5 lemons for 3 cents, how many lemons can be bought for 12 cents?

6. If 4-fifths of a yard of calico cost 8 cents, how many yards can be purchased for 25 cents?

7. If 1 and 1-fourth tuns of hay cost $8, what is the price of 1 tun?

8. If 6-sevenths of a yard of cloth cost $4, how many yards can be purchased for $12?

9. A jockey, by selling a horse for $45, gained 1-eighth of the cost: what was the cost?

10. If 1 and 1-third yards of cloth cost $5, how much can be purchased for $12?

11. By selling 5 yards of cloth for $12, I gained 1-third of the cost: what did I pay per yard?

12. By selling 7 yards of cloth for $21, I made 2-fifths of the cost: I paid for it with wheat, at $⅔ per bushel: how many bushels did I give?

13. If 3-fifths of an apple cost 2-thirds of a cent, what will 3 apples cost?

14°. What will be the cost of 11 yards of cloth, if 5 and 1-half yards cost $4⅖?

15. By selling cloth at $8 a yard, 1-fifth of the cost was lost: what part would have been gained, if the cloth had been sold for $11?

SECTION XXIII.—TABLES.

UNITED STATES MONEY.

10 mills, marked m., .. make 1 cent, .. marked ct.
10 cents .... " 1 dime, . " d.
10 dimes or 100 cents. " 1 dollar, . " $.
10 dollars .... " 1 eagle, . " E.

1. Repeat the table of United States Money.
2. How many cents in a half dime? in a quarter of a dollar? in a half dollar?
3. In 1 cent how many mills? in 2? in 3?
4. In 1 dime how many cents? in 2? in 4?
5. In 1 dollar how many cents? in 2? in 4?
6. In 20 mills how many cents? in 30? in 50?
7. In 20 cents how many dimes? in 25? in 30?

8. At 20 cents a yard, what will 3 and 3-fourths yards of calico cost? how many dimes?

9. At 25 cents a yard, what will 3 and 1-third yards of muslin cost? how many dimes?

ENGLISH MONEY.

4 farthings (far.) . make 1 penny, .. marked d.
12 pence ..... " 1 shilling, . " s.
20 shillings .... " 1 pound, .. " £.

1. Repeat the table of English Money.
2. In 1 penny how many farthings? in 2? 3? 5?
3. In 1 shilling how many pence? in 2? 3? 4?
4. How many shillings in 1 pound? in 2? 3? 4?
5. How many pence in 8 farthings? in 12? 16?
6. How many shillings in 24 pence? in 36? 56?
7. How many pounds in 40 shillings? in 60? 72?

TROY WEIGHT.

24 grains (gr.) make 1 pennyweight, . marked pwt.
20 pennyweights " 1 ounce, ... " oz.
12 ounces . . . " 1 pound, ... " lb.

1. Repeat the table of Troy Weight.
2. In 1 pennyweight how many grains? in 2? 3?
3. In 1 ounce how many pennyweights? in 2? 3?
4. In 1 pound how many ounces? in 3? 5?
5. In 24 oz. how many lb.? in 36? 48? 28? 56?

APOTHECARIES WEIGHT.

20 grains (gr.) . . make 1 scruple, . . marked ℈.
3 scruples. . . . " 1 dram, . . . " ʒ.
8 drams . . . . " 1 ounce, . . . " ℥.
12 ounces . . . . " 1 pound, . . . " ℔.

1. Repeat the table of Apothecaries Weight.
2. In 1 scruple how many grains? in 2? 3?
3. In 1 dram how many scruples? in 3? 7?
4. In 1 ounce how many drams? in 2? 3?
5. In 1 pound how many ounces? in 3? 7?
6. In 15 oz. how many lb.? in 24? in 35? 48? 50?

AVOIRDUPOIS WEIGHT.

16 drams (dr.) . . . make 1 ounce, . . marked oz.
16 ounces, ... " 1 pound, . . " lb.
25 pounds, ... " 1 quarter, . " qr.
4 quarters or 100 lb. " 1 hundred weight, cwt.
20 hundred weight, " 1 tun, . . . " T.

1. Repeat the table of Avoirdupois Weight.
2. In 2 pounds how many ounces? in 3? in 4?

3. In 2 quarters how many pounds? in 3? in 4?
4. In 2 tuns how many cwt.? in 3? in 4? in 5?
5. At 2 cents an ounce, what will 2 lb. cost?
6. At 2 dollars a quarter, what will 2 cwt. cost?

DRY MEASURE.

2 pints (pt.) . . . make 1 quart, . . . marked qt.
8 quarts . . . . . " 1 peck, . . . " pk.
4 pecks . . . . . " 1 bushel, . . " bu.

1. Repeat the table of Dry Measure.
2. In 1 quart how many pints? in 2? in 3? 5?
3. In 1 peck how many quarts? in 2? in 4? 7?
4. In 1 bushel how many pecks? in 2? in 3? 5?
5. In 2 pints how many quarts? in 5? in 8? 9?
6. In 8 quarts how many pecks? in 16? 24? 32?
7. In 4 pecks how many bushels? in 12? in 20?
8. At 5 cents a pt., what will 2 qt. of beans cost?
9. At 6 cents a qt., what will 1 peck of corn cost?
10. At 3 cts. a qt., what cost 3 and 3-fourths pk. of salt?

LIQUID OR WINE MEASURE.

4 gills (gi.) . . . make 1 pint, . . . marked pt.
2 pints . . . . . " 1 quart, . . " qt.
4 quarts . . . . " 1 gallon, . . " gal.
31½ gallons . . . . " 1 barrel, . . " bl.
63 gallons . . . . " 1 hogshead, . " hhd.
4 hogsheads . . . " 1 tun, . . . " T.

1. Repeat the table of Wine Measure.
2. In 1 pint how many gills? in 2? in 3? in 5?
3. In 2 quarts how many pints? in 3? 5? 6? 8?

4. In 1 gal. how many quarts? in 2? 3? 7? 10?
5. In 1 tun how many hogsheads? in 3? in 5? 8?
6. In 4 gi. how many pints? in 7? in 8? in 9?
7. In 3 pt. how many quarts? in 6? in 8? 11?
8. In 4 qt. how many gallons? in 7? in 11? 12?
9. At 12 cents for one pint, what will 1 and 3-fourths gallons of Port wine cost?

LONG MEASURE.

12	inches (in.). .	make	1 foot, . . .		marked	ft.
3	feet	"	1 yard, . .		"	yd.
$5\frac{1}{2}$	yards or $16\frac{1}{2}$ ft.	"	1 rod (or pole),		"	rd.
40	rods or 220 yd.	"	1 furlong, .		"	fur.
8	furlongs . . .	"	1 mile, . . .		"	mi.

1. Repeat the table of Long Measure.
2. In 2 feet how many inches? in 3? in 5? in 8?
3. In 2 yards how many feet? in 3? in 4? in 5?
4. In 2 rods how many yards? in 3? in 4? in 5?
5. In 1 furlong how many rods? how many yards?
6. In 1 mile how many furlongs? in 3? 4? 5?
7. How many inches in 8 and 5-sixths feet?
8. In 2 rods how many feet? in 3 rods? 4 rods? 5 rods? 6 rods?

SQUARE MEASURE.

144	square inches .	make	1 square foot,	marked	sq. ft.
9	square feet . .	"	1 square yard,.	"	sq. yd.
$30\frac{1}{4}$	square yards .	"	1 square rod, .	"	sq. rd.
40	square rods. .	"	1 rood, ...	"	R.
4	roods	"	1 acre, ...	"	A.
640	acres.	"	1 square mile,.	"	sq. mi.

TABLES.

1. Repeat the table of Square Measure.
2. How many square feet in 3 square yards? in 5?
3. How many roods in 4 acres? in 6? in 8? in 10?

CLOTH MEASURE.

4 nails (na.). .	make	1 quarter, . .	marked	qr.
4 quarters . . .	"	1 yard, ...	"	yd.
3 quarters . . .	"	1 ell Flemish, .	"	E. Fl.
5 quarters . . .	"	1 ell English, .	"	E. En.
6 quarters . . .	"	1 ell French, .	"	E. Fr.

1. Repeat the table of Cloth Measure.
2. How many quarters in 2 yards? in 3? 4? 5?
3. How many quarters in 1 ell Flemish? in 4? 5?
4. How many quarters in 1 ell English? in 3? 4?
5. How many quarters in 1 ell French? in 2? 4?
6. How much longer is 1 ell French than 1 ell English? than 1 ell Flemish?
7. There are 36 inches in 1 yard: how many inches are there in 1 quarter? in 1 nail?

TIME MEASURE.

60 seconds (sec.)	make	1 minute,	marked	min.
60 minutes	"	1 hour, .	"	hr.
24 hours	"	1 day, . .	"	da.
365 days 6 hours (365¼ days)	"	1 year, .	"	yr.
100 years	"	1 century,	"	cen.
7 days	"	1 week, .	"	wk.
4 weeks	"	1 month,	"	mon.
12 calendar months	"	1 year, .	"	yr.

NOTE.—One Solar year contains 365 days, 5 hours, 48 minutes, and 48 seconds, or 365¼ days *nearly*.

The following table shows the names of the different months of the year, and the number of days embraced in each.

January,	1st	mon.,	31 da.	July,	7th mon.,	31 da.
February,	2d	"	28 "	August,	8th "	31 "
March,	3d	"	31 "	September,	9th "	30 "
April,	4th	"	30 "	October,	10th "	31 "
May,	5th	"	31 "	November,	11th "	30 "
June,	6th	"	30 "	December,	12th "	31 "

The number of days in each month of the year may be retained in the mind by committing the following lines to memory:

> Thirty days has September,
> April, June, and November;
> Other months have thirty-one,
> Except the second month alone;
> To this we twenty-eight assign,
> Till leap-year gives it twenty-nine.

1. Repeat the table. What make minutes? days?

2. In 14 days how many weeks? in 21? in 28?

3. How many minutes in 1 hour and a quarter?

4. There are 12 months in 1 year: what part of a year is 1 month? what part are 2 months? 3 months? 4 months? 5? 6? 8? 9? 10?

5. If there are 30 days in 1 month, what part of a month are 3 days? what part are 5 days? 6 days? 10 days? 12 days? 15 days? 18 days? 20 days? 24 days? 27 days?

GENERAL REVIEW.

APPLICATIONS OF MENTAL ARITHMETIC,

For Advanced Classes.

SECTION XXIV.

The preceding sections contain all the elementary forms of Analysis. Those who are properly acquainted with them, will find but little difficulty in these applications.

LESSON I.—ADDITION.

SUGGESTION.—Where the numbers are large, it is better to add by 10's or 100's, as the case may be.

To add 25 and 35: 20 and 30 are 50; 5 and 5 are 10; 50 and 10 are 60. Or, 25 and 30 are 55; 55 and 5 are 60.

OF THE SIGNS.

The sign $+$, called *plus*, means more. The numbers between which it is placed are to be added.

Thus, $2 + 4$, (2 plus 4), shows that 2 and 4 are to be added together.

The sign of equality, $=$, shows that the number of units on the right and left of it are equal to each other.

Thus, $2 + 4 = 6$, means that 2 added to 4 equal 6; and is read, 2 and 4 are 6, or 2 plus 4 equal 6.

EXAMPLES. $4 + 2 =$ how many?
$3 + 5 =$ how many?
$2 + 3 + 4 =$ how many?
$1 + 3 + 5 + 7 =$ how many?
$\$4 + \$3 + \$1 + \$2 =$ how many \$'s?

1. What is the sum of $4 + 5 + 6 + 7$?
2. What is the sum of $9 + 10 + 11 + 8$?
3. What is the sum of $20 + 12 + 9 + 11$?
4. What is the sum of $24 + 20 + 12 + 30$?
5. What is the sum of $35 + 40 + 15 + 20$?
6. What is the sum of $50 + 60 + 70 + 80$?
7. What is the sum of $54 + 20 + 13 + 12$?
8. What is the sum of $\$21 + \$16 + \$13 + \$20$?

9. Bought at one time 33 gal. of oil, at another 20, at another 40, at another 50, and at another 62: how many gal. did I buy?

10. A lady paid $23 for a dress, $18 for a shawl, and $9 for a bonnet: how much did she spend?

11. I owe A $50, B $75, C $40, and D $20: how much money do I owe?

12. I collected of one man $110, of another $90, of another $75, and of another $50: how much in all?

13. Thirty-two plus $16 + 20 + 21 + 18 =$ what?

14. Fifty-nine plus $21 + 32 + 15 + 11 =$ what?

LESSON II.—SUBTRACTION.

The sign —, called *minus*, means less. Placed between two numbers, it shows that the one on the right, is to be taken (subtracted), from the one on the left.

Thus, $4 - 2 = 2$; read, 4 minus (less) 2 equal 2.

EXAMPLES. $5 - 2 =$ how many?
$3 + 9 - 8 =$ how many?
$1 + 5 + 9 - 10 =$ how many?
$2 + 3 + 4 + 8 - 11 =$ how many?
$\$5 + \$9 + \$7 + \$4 - \$12 =$ how many $'s?

REVIEW.—MULTIPLICATION.

1. What number equals 75 less 40? 160 — 120? 100 — 45? 110 — 90? 120 — 95?

2. A boy having 75 ct., purchased 55 ct. worth of goods: how much change did he receive?

3. Having $92, I purchased a watch for $73: how much had I left?

4. Bought a horse for $110, and sold him for $145: how much did I make?

5. George bought candles for 25 ct., soap for 10 ct., sugar for 35 ct., and starch for 3 ct.: he gave $1, and received 30 ct. change: was this correct?

6. A boy had $5, from which he took at one time $1 and 50 ct.; at another, 40 ct.; at another, $1 and 10 ct.: how much had he left?

7. Ten plus 22 + 19 less 8 + 3 + 9 + 6 = what?
8. 42 + 19 + 13 + 15 — 12 — 17 — 20 — 4 = ?
9. $37 + $33 + $45 + $25 — $35 — $20 — $40 = ?
10. $125 + $140 + $20 — $100 — $50 — $8 — $30 = ?
11. $160 + $80 + 130 — $210 — $30 — $10 — $40 = ?

LESSON III.—MULTIPLICATION.

The sign ×, denotes multiplication, read, *multiplied by*. When placed between two numbers, it shows that they are to be multiplied together.

Thus, 3 × 5, means 3 *multiplied by* 5.

EXAMPLES.   3 × 5 = how many?
2 × 5 × 7 = how many?
2 × 4 × 6 × 3 = how many?
$8 + $12 — $7 × 10 = how many $'s?

1. What is the product of 2+6×2? 25×2? 16×3? 20×5? 22×6? 40×5? 38×3? 60×4? 45×5? 24×6? 53×9? 65×8?

2. What is the product of 14×6? 4×7×5? 5×6×7? 9×10×5? 6×8×5?

3. What will be the cost of 5 yd. of cloth, at $2 and 50 ct. a yd.?

4. A man traveling at the rate of 5 mi. an hr., meets a stage going at the rate of 9 mi. an hr.: how far from the man will the stage be in 10 hr.?

5. What cost 9 boxes, at 7 dimes each?

6. What cost 8 lb. and 4 oz. of sugar at 12 ct. a lb.?

7. What cost 75 ft. of lumber, at 3 ct. a ft.?

8. If 1 A. of land produce 85 bu. of corn, how many bu. will 11 A. produce?

9. Bought 15 lb. of coffee at 10 ct. a lb., and 13 lb. at 9 ct. a lb.: what did the whole cost?

10. Henry has 19 ct.; George 3 times as many, lacking 10: how many have both?

11. How many yd. in 3 bales of cloth, each containing 6 pieces of 35 yd. each?

LESSON IV.—DIVISION.

The sign ÷, is read, *divided by*. When placed between two numbers, it shows that the first is to be divided by the second.

Thus, 6÷2, means that 6 is to be divided by 2.

EXAMPLES.

6 ÷ 2 = how many?

4 × 5 ÷ 2 = how many?

4 + 10 × 2 ÷ 7 = how many?

$3 + $12 − $5 × 4 ÷ 8 = how many $'s?

REVIEW.—DIVISION.

1. How often is 3 contained in 48? in 51? in 60? in 75? in 81? in 90? in 144?

2. Divide 125 by 5; multiply the result by 10; then divide by 2: what is the last quotient?

3. Multiply 14 by 20, and divide the product by 7.

4. What does $12 \times 13 \div 2 \div 6 = $?

5. What does $15 \times 12 \div 3 \div 12 - 3 = $?

6. What does $27 + 9 \div 12 + 20 - 17 \div 2 \times 5 = $?

7. What is 1-half of $28 + $ 1-third of 72?

8. What are 9-fifteenths of $120 + \frac{2}{3}$ of 60?

9. If a boat sail 48 mi. in 12 hr., how far will she sail in 6 hr.?

10. At 15 ct. a lb., what quantity of beef can be purchased for $6?

11. Three men bought a horse for $90: after keeping him 6 wk., at $3 a wk., they sold him for $99: what did each man lose?

12. Seven multiplied by 9, divided by 3, 2 added, 13 subtracted, and divided by 5, will = what?

13. Add 10 to $12 \times 3 \div 6 + 5 \div 8 + 10 \times 5 \div 6 \div 2 = $ what?

14. Seventeen $+ 6 - 8 \div 3 \times 8 - 6 + 14 \div 4 \times 8 - 8 + 12 \div 10 + 6 + 7 - 12 \times 6 = $ what?

15. What number added to itself will give a sum equal to 14?

Explanation.—If a number be added to itself, the sum will be 2 times the number: 14, then, is 2 times what number?

16. What number added to itself 3 times, will make 32?

17. Divide 16 into 2 parts, so that the second part will be 3 times the first.

Explanation.—The *sum* of the parts (16), will be 4 times the first part.

18. Divide 48 into 2 such parts that the second shall be 7 times the first. 48 is 8 times what number?

19. Divide 24 into 3 parts, so that the second shall be 2 times the first, and the third 3 times.

20. A boy being asked the number of ct. he had, replied: "Five times the number I have, is just 40 less than 10 times the number:" how many had he?

21. Find a number which, being multiplied by $2 \div 8 \times 3 - 9 \div 3 \times 3 + 11 \times 3 - 1 \div 10$, equals 5.

Explanation.—Begin with the last number mentioned, and reverse every operation indicated by the signs: thus, $5 \times 10 = 50$; $50 + 1 = 51$; $51 \div 3 = 17$, &c.

22. What does $12 \times 5 + 3 \div 7 + 11 \div 5 - 1 + 10 + 14 \div 3 + 19 + 8 \div 9 + 17 + 8 = ?$

23. What does $13 + 27 + 14 + 10 \div 8 + 21 + 13 \div 7 + 14 + 20 + 23 + 3 \div 11 = ?$

24. What does $19 + 2 - 13 \times 6 \div 4 + 7 - 12 \times 5 \div 7 + 15 - 11 \times 8 \div 12 + 15 - 14 = ?$

LESSON V.—PRINCIPLES.

1. When 10 was taken from a number, only 2-thirds of the number remained: what was the number?

2. The sum of two numbers is 12; if 6 be added to the sum, the result will be twice the greater number: what are the numbers?

3. The sum of two numbers is 16 more than their difference: if their difference be 4, and 8 one of the numbers, what is the other number?

4°. The sum of two numbers diminished by the less gives 15: if 10 is the less number, what is their difference?

REVIEW.—PRINCIPLES.

5. If 6 be taken from the difference of two numbers, the remainder will be 2: if 4 is one of the numbers, what is the other?

6. If 10 be added to the difference of two numbers, the sum will be 6 more than the greater number, which is 19: what is the less number?

7. If 10 be taken from the sum of two numbers, of which 5 is one, there will be 8 left: what is the other number?

8. By what part of 6 must 4 be × to = $\frac{3}{5}$ of 20?

9. What number ÷ 8 will give 13 for a quotient?

10. What number × 12, will give 156 for a product?

11. If 15 be multiplied by some number and 20 added to the product, the sum will be 200: what is the multiplier?

12. A certain number × 2, gives a result as much less than 20 as the number is greater than 7; but when it is subtracted from 11, it leaves the same remainder as 5 from 7: what is the number? why?

13. Six is contained in a certain number 12 times, with a remainder of 5: what is the number?

14. If 12 be added to a certain number, 7 will be contained 9 times in the sum, with a remainder of 1: what is the number?

15. If 13 be taken from a certain number, 8 will be contained 10 times in the difference, with a remainder of 3: what is the number?

16. When the divisor of 132 was increased by 6, the quotient was found to be 11: what was the divisor?

17. When the divisor of 72 was multiplied by 2, the quotient was 9: what was the divisor?

18. When the divisor of 84 was divided by 3, the quotient was 4: what was the divisor?

19. If 1 be added to the number of times a certain number is contained in 60, the result will be 11: what is that number?

20. Twice the greater of two numbers, — 2 = their sum, which is 20: what are the numbers?

21. A boy received of his father 3 ct.; of his mother twice as many less 1; if he had received from his father 5 ct. more, his father would have given him 4 times as many as his sister: how many ct. did he receive?

SECTION XXV.—QUESTIONS.

LESSON I.

1. If $\frac{1}{3}$ of a yd. of cloth cost $2, what cost $\frac{1}{4}$ of a yd.?
2. If $\frac{2}{3}$ of a yd. of cloth cost $5, what cost $\frac{3}{4}$ of a yd.?

ANALYSIS.—1-*third of a yd. will cost* 1-*half as much as* 2-*thirds; and* 3-*thirds, or a whole yd., will cost* 3 *times as much as* 1-*third; and,*

One-*fourth of a yd. will cost* 1-*fourth as much as* 1 yd., *and* 3-*fourths,* 3 *times as much as* 1-*fourth.*

3. If $\frac{2}{5}$ of a bl. of flour cost $3, what cost $\frac{2}{3}$ of a bl.?
4. If $\frac{4}{7}$ of a yd. of muslin cost 24 ct., what will $\frac{5}{14}$ of a yd. cost?
5. If $\frac{5}{9}$ of a tun of hay cost $15, what will one-half a tun cost?
6. If $\frac{3}{8}$ of an orchard contain 30 fruit trees, how many trees are there in $\frac{7}{18}$ of it?
7. If $1\frac{2}{5}$ yd. of cloth cost $14, what cost $2\frac{1}{2}$ yd.?
8. If $1\frac{1}{2}$ bl. of flour cost $5\frac{1}{4}$, what cost $2\frac{1}{2}$ bl.?
9. If $3\frac{1}{3}$ lb. of cheese cost 20 ct., what cost $2\frac{5}{8}$ lb.?

10. A traveled 30 mi. in $3\frac{3}{4}$ hr.: at that rate, how far can he travel in $7\frac{1}{4}$ hr.?

11. If a man earn $\$1\frac{1}{4}$ in 10 hr., how much can he earn in 11 hr.?

12. A can earn $\$9\frac{3}{5}$ in 6 da., of 8 hr. each: how much can he earn in 7 da., of 9 hr. each?

13. If $5\frac{3}{4}$ bu. of wheat cost $\$9\frac{1}{5}$, what cost $3\frac{2}{3}$ bu.?

14. If $8\frac{1}{3}$ is $\frac{5}{7}$ of a number, what is $\frac{4}{5}$ of it?

15. If $3\frac{1}{2}$ is $2\frac{2}{3}$ times some number, what is $2\frac{1}{2}$ times that number?

16. If $\frac{3}{4}$ of a bl. of flour cost $\$4\frac{1}{2}$, what cost $\frac{2}{3}$ of a bl.?

17. If $\frac{2}{3}$ of a yd. of lace cost $\$\frac{3}{5}$, what cost $\frac{5}{6}$ of a yd.?

18. If the wages of 3 men for 5 da. is $30, what will be the wages of 4 men for 7 da.?

SUGGESTION.—First find the wages of *one* man for *one* day.

19. If 6 persons spend $36 in 8 da., how much, at that rate, would 5 men spend in 12 da.?

20. If 3 men can build 12 rd. of wall in 8 da., how many rd. can 5 men build in 3 da.?

21. If 6 horses eat 36 bu. of oats in 10 da., how many bu. will 5 horses eat in 9 da.?

22. If 5 oxen eat 2 A. of grass in 6 da., in how many da. will 12 oxen eat 8 A.?

23. If a family of 8 persons spend $400 in 5 mon., how much would maintain them 8 mon., if 3 more persons were added?

ANALYSIS.—$400 *for* 5 *mon. is* $80 *for* 1 *mon.; if* 8 *persons spend* $80 *in* 1 *mon.*, 1 *person spends* $10 *in* 1 *mon. Hence,* 11 *persons spend* $110 *in* 1 *mon., and* $880 *in* 8 *mon.*

24. If 10 oxen can be kept on 5 A. for 3 mon., how many sheep can be kept on 15 A. for 5 mon., if 7 sheep eat as much as 1 ox?

LESSON II.

1. If 5 men can do a piece of work in 18 da., how many men will do it in 9 da.?

2. If 8 men can do a piece of work in 15 da., how many men can do it in 12 da.?

3. If 9 pipes fill a cistern in $2\frac{1}{2}$ hr., in what time will 5 such pipes fill it?

4. If 5 men do a piece of work in 6 da., how many can do a piece twice as large, in 1-fifth the time?

5. If 8 men can do a piece of work in 5 da., in what time can 5 men do it?

6. If 6 men can do a piece of work in 5 da., in what time can they do it, if they receive the assistance of 3 additional men when the work is half completed?

7. If 7 men can do a piece of work in 4 da., in what time can it be done, if 3 of the men leave when the work is half completed?

8. If 20 lb. of flour afford 8 five ct. loaves, how many one ct. loaves will it furnish? how many four ct. loaves? how many ten ct. loaves?

9. If 10 lb. of flour afford 6 five ct. loaves, how many 3 ct. loaves will it furnish?

ANALYSIS.—*It will afford* 5 *times as many* 1 *ct. loaves as* 5 *ct. loaves:* 1-*third as many* 3 *ct. loaves as* 1 *ct. loaves.*

10. If a sack of flour make 20 three ct. loaves, how many 4 ct. loaves will it make? 5 ct. loaves?

11. If the 5 ct. loaf weigh 8 oz. when flour is $3 a bl., what should it weigh when flour is $1 a bl.? what if flour is $2 a bl.? $4 a bl.?

12. A 4 ct. loaf weighs 10 oz. when flour is $6 a bl.; what will it weigh when flour is $5 a bl.?

Analysis.—*If flour were $1 a bl., it ought to weigh 6 times as much as when flour is $6 a bl.; that is, 60 oz.: and,*

When flour is $5 a bl., it ought to weigh ⅕ as much as when it is $1 a bl., that is, 12 oz.

13. If the 3 ct. loaf weigh 7 oz. when flour is $3½ a bl., what ought it to weigh when flour is $2½ a bl.?

14. If 6 men can mow a field in 5½ da., how much time would be saved by employing 4 more men?

LESSON III.

1. A and B hired a pasture for $45: A pastured 4 cows, and B 5 cows: what should each pay?

Analysis.—*They together pastured 9 cows, of which 4/9 were A's, and 5/9 B's; hence, A should pay 4/9 of $45, which are $20; and B 5/9 of $45, which are $25.*

2. William had 3 ct., Thomas 4 ct., and John 5 ct.; they bought 36 peaches: what was the share of each?

3. Two men paid $3 for 7½ dozen oysters: the first paid $2, and the second $1: how many should each have?

4. A and B bought a horse for $40; A paid $25, and B the rest: they sold him for $56: what should each receive?

5. A boat worth $860, of which ⅛ belonged to A, ¼ to B, and the rest to C, was entirely lost: what loss will each sustain, it having been insured for $500?

6. Two boys bought a silver watch for $7: the first paid $2½; the second, $4½; they sold it for $21: what was each one's share?

7. A man failing, paid 80 ct. on each dollar of his indebtedness: what did I receive, if he owed me $60?

Analysis.—*80 ct. are ⅘ of $1; he therefore paid me ⅘ of $60, or $48.*

8. A grocer failing, pays 60 ct. on the dollar: what will B receive to whom he owes $25?

9. A trader failing, pays only 15 ct. on the dollar: what will C receive to whom he owes $80?

10. A and B rent a pasture for $25: A puts in 27 oxen, and B 180 sheep: what should each pay, supposing an ox to eat as much as 10 sheep?

11. A and B rent a pasture for $60: A puts in 14 horses, and B 15 cows: what should each pay, if 2 horses eat as much as 3 cows?

12. A and B rent a pasture for $75: A puts in 8 horses; B 15 oxen and 120 sheep: what should each pay, if a horse eat as much as 20 sheep, and 2 horses as much as 3 oxen?

13. A and B rent a pasture for $35; A puts in 4 horses 2 wk.; B, 3 horses 4 wk.: what ought each to pay?

ANALYSIS.—4 horses for 2 wk. = 1 horse for 8 wk.; and 3 horses for 4 wk. = 1 horse for 12 wk.: 8 wk. and 12 wk. are 20 wk; hence,

A must pay $\frac{8}{20}$, or $\frac{2}{5}$ of the rent, = $14; and B $\frac{12}{20}$, or $\frac{3}{5}$ of the rent, = $21.

14. C and D join their stocks in trade; C puts in $50 for 4 mon., and D $60 for 5 mon.: they gain $45: what is the share of each?

15. Two masons, A and B, built a wall for $81; A sent 3 men for 4 da., and B 5 men for 3 da.: what ought each to receive?

16. A and B traded in company; A put in $2 as often as B put in $3; A's money was employed 5 mon., and B's 4 mon.; they gained $55: what was each man's share?

17. E and F rented a field for $27; E put in 4 horses for 5 mon., and F 10 cows for 6 mon.: what ought each to pay, if 2 horses eat as much as 3 cows?

LESSON IV.

1. Divide 20 apples between A and B, so that A may get 2 as often as B gets 3.

ANALYSIS.—*Of each 5 apples, A must get 2 and B 3. In 20 apples there are 4 times 5 apples; hence,*
A must get 4 times 2 apples, or 8 apples; and B 4 times 3 apples, or 12 apples.

2. Divide 28 ct. between John and James, so that John may get 3 as often as James gets 4.

3. Divide 45 ct. between A, B, and C, so that A may get 4 ct. as often as B gets 3, and C gets 2.

4. In an orchard of 96 trees, there are 5 apple-trees for 3 peach-trees: how many of each kind?

5. On a farm there are 60 animals—horses, cows, and sheep; for each horse there are 3 cows, and for each cow there are 2 sheep: how many animals of each kind?

6. A school of 35 pupils has 2 boys for 3 girls: how many of each in the school?

7. What number is that which being added to 3 times itself will make 48?

8. Divide 42 plums between A, B, and C, so that B may get twice, and C three times as many as A.

9. Mary has 25 yd. of ribbon: she wishes to divide it into two parts, so that one shall be 4 times the length of the other: what will be the length of each part?

10. Divide 35 cherries between Emma, Agnes, and Sarah, so that Agnes shall have twice as many as Emma, and Sarah twice as many as Agnes.

11. Divide 28 into two parts, so that one shall be 3 times a certain number, and the other 4 times.

LESSON V.

1. What part of 8 is 2? what part is 4? is 1?

2. How many times does 10 contain 2? 2 is what part of 10?

3. Twelve is how many times 2? 2 is what part of 12? what is the ratio of 2 to 12?

EXPLANATION.—When two numbers are compared, to see how many times greater one is than the other, what do you find? *Ans. The ratio of the numbers.*

How do you find the ratio or relation of two numbers? *Ans. Divide the second number by the first.*

4. How many times does 18 contain 9? what is the ratio of 9 to 18?

5. What is the ratio of 12 to 36? 9 to 45? 11 to 66? 13 to 52? 2 to 1? 4 to 3?

6. What is the ratio of $2\tfrac{1}{2}$ to 5? $6\tfrac{1}{4}$ to $12\tfrac{1}{2}$? $\tfrac{1}{4}$ to $\tfrac{1}{2}$? of $\tfrac{2}{3}$ to $\tfrac{5}{8}$? $\tfrac{2}{3}$ to $\tfrac{4}{5}$? $\tfrac{1}{2}$ to $\tfrac{1}{3}$?

7. If the ratio of two numbers is 5, and 6 is the less number, what is the greater?

8. The ratio of 7 to 21 is equal to the ratio of some number to 36: what is the number?

9. Five less than the ratio of 2 to 20 is $\tfrac{1}{4}$ of the ratio of 2 to what?

10. The ratio of 2 to 18, plus 3, is 7 less than the ratio of 2 to what?

11. The ratio of 9 to 27, increased by 5, is equal to the ratio of $2\tfrac{1}{3}$ to what?

12. Divide 25 ct. between John and George, so that their shares shall be in the ratio of 2 to 3.

Explanation.—When two numbers are in the ratio of 2 to 3, one will contain 2 as often as the other contains 3.

13. Divide the number 48 into two parts that shall be in the ratio of 5 to 7.

14. Divide the number 60 into three parts that shall be to each other as 3, 4, and 5.

15. Divide the number 70 into 4 parts that shall be to each other as 1, 2, 3, and 4.

16. Divide the number 22 into two parts that shall be to each other as $2\tfrac{1}{2}$ to 3.

ANALYSIS.—$2\tfrac{1}{2}$ and 3 are $5\tfrac{1}{2}$ *units. The first part will therefore be as many times* $2\tfrac{1}{2}$, *and the second as many times* 3, *as* $5\tfrac{1}{2}$ *are contained times in* 22.

17. Divide 16 apples between Henry and Oliver, so that their shares shall be in the ratio of $1\tfrac{1}{2}$ to $2\tfrac{1}{2}$.

18. Divide the number 39 into three parts that shall be to each other as $\tfrac{1}{2}$, $\tfrac{1}{3}$, and $\tfrac{1}{4}$.

19. Divide 14 ct. between A and B, so that B may have $1\tfrac{1}{3}$ times as many as A.

ANALYSIS.—*As often as A gets* 1 *ct., B gets* $1\tfrac{1}{3}$; *that is, of each* $2\tfrac{1}{3}$ *ct. A gets* 1, *and B* $1\tfrac{1}{3}$ *ct.; hence,*
A gets as many times 1 *ct. and B as many times* $1\tfrac{1}{3}$ *ct., as* $2\tfrac{1}{3}$ *ct. are contained times in* 14 *ct.* Ans. *A* 6 *ct., B* 8 *ct.*

20. John and James together have 33 marbles; James has $1\tfrac{3}{4}$ times as many as John: how many has each?

21. William's age is $1\tfrac{2}{3}$ times Frank's age; the sum of their ages is 32 yr.: what is the age of each?

22. A basket contains 30 apples: the number of those which are sound, is $2\tfrac{1}{3}$ times the number of those not sound: how many are there of each?

23. Two men built 27 ft. of wall: how much did each build, if one built $\tfrac{4}{5}$ as much as the other?

24. A, B, and C, have $42; B has half as many as A, and C half as many as B: how many has each?

25. On a farm there are 104 animals—hogs, sheep, and cows; there are $\frac{2}{3}$ as many sheep as hogs, and $\frac{3}{4}$ as many cows as sheep: how many are there of each?

LESSON VI.

1. Divide 15 ct. between A and B, so that B may have 3 more than A.

ANALYSIS.—*Reserving 3 ct. for the number that B receives more than A, there are 12 ct. left; dividing these equally, A will get 6 ct.; and B, 6 ct. plus 3 ct. reserved, or 9 ct.*

2. Thomas has 5 apples more than James, and they both together have 19: how many has each?

3. The sum of two numbers is 31, and the greater exceeds the less by 7: what are the numbers?

4. Thomas and James each had the same number of ct., when Thomas found 8 ct. more; they then had together 32 ct.: how many had each?

5. Thomas and William each bought the same number of peaches; after Thomas ate 4, and William 6, they both together had 20 left: how many had each remaining?

6. Mary bought twice as many cherries as Sarah; and after Mary ate 7, and Sarah 5, they had only 24 left: how many had each left?

7. If 5 be added to the treble of a certain number, the sum will be 50: what is the number?

8. If $\frac{3}{4}$ of a certain number be increased by 10, the sum will be 31: what is the number?

9. If $\frac{4}{5}$ of a number be diminished by 7, the remainder will be 21: what is the number?

10. James is 4 yr. older than Henry, and Henry is 2 yr. younger than Oliver; the sum of their ages is 37 yr.: what is the age of each?

11. Mary has 8 ct. more than Jane, and Sarah 3 less than Mary; they all have 43 ct.: how many has each?

12. The sum of the ages of Mary and Frank is 42 yr.; Mary is twice as old as Frank, less 3 yr.: what is the age of each?

13. I bought a watch, a chain, and a ring, for $62; the chain cost $5 less than the ring, and the watch $12 more than the chain: what did I pay for each?

14. Thirty ct. are 6 ct. less than $\frac{1}{2}$ of $\frac{4}{5}$ of my money: how much have I?

15. John has twice as much money as James, $+$ $3; Frank has as much as John and James, $+$ $7; together they have $55: how much has each?

LESSON VII.

1. Divide the number 15 into two parts, so that the less part may be $\frac{2}{3}$ of the greater.

ANALYSIS.—*The greater part being 3-thirds, and the less, 2-thirds, their sum will be 5-thirds: 15, then, is 5-thirds of what number?*

2. Thomas and John have $60 to pay; John has $\frac{2}{3}$ as much to pay as Thomas: what must each pay?

3. The time past from noon is equal to half the time to midnight: what o'clock is it?

4. The time elapsed since noon is $\frac{3}{5}$ of the time to midnight: what is the hour?

5. I had 56 mi. to travel in 2 da.; the second da., I went $\frac{3}{4}$ as far as the first: how far did I travel each da.?

6. Divide 100 into two such parts, that $\frac{5}{7}$ of the first less 8 will = the second.

7. Divide the number 45 into three such parts, that the second shall be $\frac{1}{2}$, and the third $\frac{3}{4}$ of the first part.

The sum of all the parts will be nine-fourths of the first part.

8. A, B, and C, together have 40 ct.; B has $\frac{3}{5}$ as many as A, and C $\frac{2}{3}$ as many as B: how many has each?

9. A tree 70 ft. long was broken into 3 pieces; the middle part was $\frac{5}{6}$ of the top part; the lower part was $\frac{3}{4}$ of the middle part: what was the length of each?

10. I bought a hat, coat, and vest, for $34; the hat cost $\frac{2}{3}$ of the price of the coat, and the vest $\frac{3}{4}$ the price of the hat: what was the cost of each?

11. In a field containing 55 sheep and cows, $\frac{1}{3}$ of the cows = $\frac{2}{7}$ of the sheep: how many are there of each?

12. The sum of two numbers is 100; and $\frac{1}{3}$ of the less equals $\frac{2}{5}$ of the greater: what are the numbers?

13. One-fourth of Mary's age = $\frac{1}{3}$ of Sarah's, and the sum of their ages is 14 yr.: what the age of each?

14. Divide 38 ct. between A and B, so that $\frac{2}{3}$ of A's share may be equal to $\frac{3}{5}$ of B's.

ANALYSIS.—*If $\frac{2}{3}$ of A's share = $\frac{3}{5}$ of B's, then $\frac{1}{3}$ of A's share = $\frac{1}{2}$ of $\frac{3}{5}$; that is, $\frac{3}{10}$ of B's; and the whole of A's = 3 times $\frac{3}{10}$, that is, $\frac{9}{10}$ of B's; and,*

Hence, $\frac{10}{10}$ of B's + $\frac{9}{10}$ of B's = $\frac{19}{10}$ of B's = 38 ct.: 38 ct., then, are $\frac{19}{10}$ of what number?

15. Divide the number 51 into two such parts, that $\frac{2}{3}$ of the first will equal $\frac{3}{4}$ of the second.

16. In an orchard of 65 apple and peach-trees, $\frac{2}{3}$ of the apple-trees = $\frac{4}{7}$ of the peach-trees: how many are there of each?

17. From C to D is 66 mi.; A left C at the same time B left D; when they met, $\frac{2}{3}$ of the distance A had traveled = $\frac{5}{9}$ of the distance B had traveled: how much farther did B travel than A?

18. The time past noon, + 3 hr., is equal to $\frac{1}{2}$ of the time to midnight: what is the hour?

REVIEW.—QUESTIONS. 139

19. What is the hour in the afternoon, when the time past noon is equal to $\frac{1}{5}$ of the time past midnight?

Explanation.—Since the time past noon is one-fifth of the whole time from midnight, the time from midnight to noon, which is 12 hr., must equal the remaining four-fifths of the time.

20. What is the hour in the afternoon, when the time past noon is $\frac{1}{4}$ of the time past midnight?

21. What is the hour of the day, when $\frac{1}{2}$ of the time past noon is $\frac{1}{20}$ of the time past midnight?

LESSON VIII.

1. What number is that to which, if its half be added, the sum will be 15?

ANALYSIS.— *The number $+$ its $\frac{1}{2} = \frac{3}{2}$ of the number. Now, if $\frac{3}{2} = 15$, $\frac{1}{2} = 5$, and $\frac{2}{2}$, or the whole number, $= 10$.*

2. What number is that to which if its $\frac{2}{3}$ be added, the sum will be 20?

3. If to Mary's age its $\frac{2}{5}$ be added, the sum will be 21 yr.: what is her age?

4. What number is that which being doubled, and increased by its $\frac{3}{5}$, the sum will be 52?

5. What number is that which being doubled, and diminished by its $\frac{4}{7}$, the remainder will be 40?

6. What number is that which being trebled, and diminished by its $\frac{3}{5}$, the remainder will be 48?

7. If to David's age you add its $\frac{1}{2}$ and its $\frac{2}{3}$, the sum will be 26: what is his age?

8. If to Sarah's age you add its $\frac{1}{3}$, its $\frac{1}{4}$, $+$ 10 yr., the sum will be twice her age: how old is she?

9. Thomas spent $\frac{2}{5}$ of his money, and had 30 ct. left: how much had he at first?

10. If to a certain number you add its $\frac{1}{2}$, its $\frac{3}{5}+27$, the number will be trebled: what is the number?

11. A father is 40 yr. older than his son; the son's age is $\frac{3}{11}$ of the father's age: what is the age of each?

12. If to Susan's age you add its $\frac{4}{5}+18$ yr., the sum will be 3 times her age: how old is she?

13. A piece of flannel, losing $\frac{2}{9}$ of its length by shrinkage, measured 28 yd.: what was its length?

14. The distance from A to B is $\frac{1}{2}$ the distance from C to D, and $\frac{2}{3}$ of the distance from A to B, $+$ 20 mi., $=$ the distance from C to D: what is the distance from A to B, and from C to D?

15. My age $+$ its $\frac{1}{3}$, its $\frac{1}{5}$, and its $\frac{5}{9}=94$: what is my age?

LESSON IX.

1. If A can do a piece of work in 2 da., what part of it can he do in 1 da.?

2. A can drink a keg of cider in 4 da.: what part of it can he drink in 1 da.?

3. B can do a piece of work in $\frac{1}{2}$ a da.: how many times the work can he do in 1 da.?

4. C can mow a certain lot in $\frac{3}{8}$ of a da.: how many such lots can he mow in a da.?

5. A can mow a certain field in $2\frac{1}{2}$ da.: what part of it can he mow in 1 da.?

ANALYSIS.—*If he can mow the field in $2\frac{1}{2}$ da., he could mow twice the field in 5 da., and $\frac{1}{5}$ of twice the field, that is, $\frac{2}{5}$ of the field in 1 da.*

6. B can dig a trench in $3\frac{1}{2}$ da.: what part of it can he dig in 1 da.?

7. C can walk from Cincinnati to Dayton in $3\frac{1}{5}$ da.: what part of the distance can he walk in 2 da.?

8. A can do $\frac{1}{2}$ of a piece of work in 1 da., and B $\frac{1}{4}$ of it: what part of the work can both do in a da.?

9. A can do $\frac{1}{2}$, B $\frac{1}{4}$, and C $\frac{1}{5}$ of a piece of work in 1 da.: what part of it can they all do in a da.?

10. If A can do a piece of work in 2 da., and B in 3 da.: in what time can they both together do it?

ANALYSIS.—*If A does it in 2 da., he can do $\frac{1}{2}$ of it in 1 da.; and if B does it in 3 da., he can do $\frac{1}{3}$ of it in 1 da.; hence,*

Both can do $\frac{1}{2} + \frac{1}{3} = \frac{5}{6}$ *in 1 da.: it will require them both as many da. as $\frac{5}{6}$ are contained times in 1,* (the whole work). *Ans.* $1\frac{1}{5}$ *da.*

11. A can dig a trench in 6 da., and B in 12 da.: in what time can they both together do it?

12. C alone can do a piece of work in 5 da., and B in 7 da.: in what time can both do it?

13. A can do a piece of work in 2 da., B in 3 da., and C in 6 da.: in what time can all three do it?

14. A cistern has 3 pipes; the first will empty it in 3 hr., the second in 5 hr., and the third in 6 hr.: in what time will all three empty it?

15. A and B mow a field in 4 da.; B can mow it alone in 12 da.: in what time can A mow it?

Explanation.—Both mow one-fourth in 1 da., and B one-twelfth in 1 da.; therefore, A can mow one-fourth less one-twelfth, which is one-sixth in 1 da.; hence, he can mow it in 6 da.

16. A man and his wife can drink a keg of beer in 12 da.; when the man is away, it lasts the woman 30 da.: in what time can the man drink it alone?

17. Three men, A, B, and C, can together reap a field of wheat in 4 da.; A can reap it alone in 8 da., and B in 12 da.: in what time can C reap it?

18. A can do a piece of work in $\frac{1}{2}$ a da., and B in $\frac{1}{3}$ of a da.: how long will it take both to do it?

ANALYSIS.—*If A does the work in $\frac{1}{2}$ of a da., he can do 2 times the work in 1 da.; and*

If B does the work in $\frac{1}{3}$ of a da., he can do 3 times the work in 1 da.; hence,

They both together can do 5 times the work in 1 da., or the whole work in $\frac{1}{5}$ of a da.

19. A cistern has two pipes; by the 1st it may be emptied in $\frac{1}{3}$ of an hr., and by the 2nd in $\frac{1}{5}$ of an hr.: in what time will it be emptied by both together?

20. A can alone dig a cellar in $2\frac{1}{2}$ da., and B in $3\frac{1}{3}$ da.: in what time can they both together dig it?

Explanation.—A digs 2-fifths in 1 da., and B 3-tenths in 1 da.; hence, they both dig 7-tenths in 1 da.

21. C can reap a field of wheat in 5 da., and D can reap it in $3\frac{1}{3}$ da.: in what time can both reap it?

22. A can do a piece of work in 3 da., and B a piece 3 times as large in 7 da.: in what time can they together do a piece 5 times as large as the piece done in 3 da.?

23. A cistern of 100 gal., is emptied in 20 min. by 3 pipes; the 1st discharges $\frac{1}{2}$ a gal. in a min.; the 2nd, $1\frac{1}{2}$ gal.: how much does the 3rd discharge?

LESSON X.

1. Two-thirds of $1\frac{1}{5}$ is $\frac{2}{7}$ of what number?

ANALYSIS.—$1\frac{1}{5}$ *is $\frac{6}{5}$; $\frac{2}{3}$ of $\frac{6}{5}$ is twice $\frac{1}{3}$ of $\frac{6}{5}$; $\frac{1}{3}$ of $\frac{6}{5}$ is $\frac{2}{5}$, and twice $\frac{2}{5}$ are $\frac{4}{5}$; if $\frac{4}{5}$ is $\frac{2}{7}$ of some number, $\frac{1}{2}$ of $\frac{4}{5}$, or $\frac{2}{5}$ is $\frac{1}{7}$ of the number; $\frac{2}{5}$ is $\frac{1}{7}$ of seven times $\frac{2}{5}$, which are $\frac{14}{5}$.*

    2. $\frac{5}{9}$ of $5\frac{2}{5}$ is $\frac{8}{9}$ of what number?
    3. $\frac{4}{7}$ of $4\frac{3}{8}$ is $\frac{5}{11}$ of what number?
    4. $\frac{5}{7}$ of $5\frac{4}{9}$ is $\frac{7}{10}$ of what number?
    5. $\frac{2}{3}$ of $2\frac{2}{5}$ is $\frac{1}{2}$ of how many times 2?

REVIEW.—QUESTIONS.

6. Three-fifths of $1\frac{1}{9}$ is $\frac{2}{7}$ of how many times 4?

7. Three-fourths of $3\frac{1}{5}$ is $\frac{3}{8}$ of how many times 3?

8. John has 10 marbles, and $\frac{4}{5}$ of what John has is the $\frac{8}{11}$ of what James has: how many has James?

9. Jane received $\frac{3}{5}$ of 60 plums; she gave away $\frac{4}{9}$ of her $\frac{3}{5}$: how many were left?

10. James has a given distance to travel; after going 35 mi., there remains $\frac{3}{7}$ of the distance: when he has gone $\frac{3}{7}$ of the remainder, how far must he then go?

11. A horse cost $40; $\frac{3}{4}$ of the price of the horse $=\frac{6}{5}$ of the price of the cart: what did the cart cost?

12. B's coat cost $27, and his hat $8; $\frac{4}{9}$ of the cost of the coat $+\frac{3}{4}$ that of the hat, $=\frac{3}{5}$ of the cost of his watch: what did the watch cost?

13. Mary lost $\frac{2}{7}$ of her plums; she gave $\frac{2}{5}$ of the remainder to Sarah, and had 6 plums left: how many had she at first?

14. John has 12 ct.; $\frac{2}{3}$ of his money $=\frac{1}{2}$ of $\frac{4}{5}$ of William's money: how much has William?

15. From A to B is 36 mi.; $\frac{5}{7}$ of this is $\frac{2}{4}$ of the distance from C to D: what is the distance from C to D?

16. On counting their money, it was found that A had 12 ct. more than B; and that $\frac{1}{2}$ of B's $=\frac{2}{7}$ of A's: how much had each?

17. In an orchard, $\frac{1}{3}$ are apple-trees, $\frac{1}{4}$ are pear-trees, $\frac{1}{12}$ are plum-trees, and the remainder, which is 32, cherry-trees: how many trees are there of each kind?

18. In an orchard of apple and pear-trees, the latter are $\frac{2}{9}$ of the whole; the apple-trees are 25 more than the pear-trees: how many are there of each?

19. In an orchard of apple, plum, and cherry-trees, 69 in all, the plum-trees $=\frac{1}{3}$ of the apple-trees, and the cherry-trees $=\frac{1}{2}$ of the apple-trees $+\frac{1}{4}$ of the plum: how many trees are there of each kind?

20. The age of Jane is $\frac{7}{8}$ of the age of Sarah, and $\frac{4}{9}$ of both their ages is $\frac{5}{3}$ of the age of Mary, which is 12 yr.: what are the ages of Jane and Sarah?

21. How many times $\frac{3}{11}$ of 55, is twice that number of which $\frac{4}{5}$ of 30 is $\frac{4}{9}$?

22. John's money is $\frac{3}{8}$ of Charles's; and $\frac{3}{4}$ of John's $+ \$33 =$ Charles's: how much has each?

LESSON XI.

1. A hare takes 4 leaps while a hound takes 3; 2 of the hound's leaps = 3 of the hare's: how many leaps must the hound take to gain the length of a hare's leap on the hare?

ANALYSIS.—*Since 2 of the hound's leaps = 3 of the hare's, 1 of the hound's leaps = $1\frac{1}{2}$ of the hare's, and 3 of the hound's leaps = $4\frac{1}{2}$ of the hare's; hence,*

In taking 3 leaps the hound gains $\frac{1}{2}$ the length of a hare's leap on the hare; therefore,

The hound must take 6 leaps to gain 1 leap on the hare.

2. Henry takes 6 steps while John takes 5; but 4 of John's steps = 5 of Henry's: how many steps must John take, to gain one of Henry's steps on him?

3. Henry is 30 steps before John, but John takes 7 steps while Henry takes 5: supposing the length of their steps to be equal, how many steps must John take to overtake Henry?

4. A hare is 10 leaps before a hound, and takes 4 leaps while the hound takes 3; but 2 of the hound's leaps = 3 of the hare's: how many leaps must the hound take to catch the hare?

ANALYSIS.—*Since the hound takes 3 leaps while the hare takes 4, the hound will take 1 leap while the hare takes $1\frac{1}{3}$ leaps, and 2 leaps, while the hare takes $2\frac{2}{3}$; but,*

REVIEW.—QUESTIONS. 145

By the second condition, 2 of the hound's leaps = 3 of the hare's; therefore,

In making 2 leaps the hound gains $\frac{1}{3}$ of a hare's leap on the hare; that is, 3 leaps — $2\frac{2}{3}$ leaps; hence,

To gain 10 of the hare's leaps, the hound must make as many times 2 leaps as $\frac{1}{3}$ is contained times in 10, that is, 60 leaps.

5. A hare is 100 leaps before a hound, and takes 5 leaps while the hound takes 3; but 3 leaps of the hound = 10 of the hare: how many leaps must the hound take to catch the hare?

6. N is 35 steps ahead of M, and takes 7 steps while M takes 5; but 4 of M's steps = 7 of N's: how many steps must M take to overtake N? How many more will N have made, when he is overtaken?

7. A hare is 8 leaps before a hound, and takes 3 leaps while the hound takes 2; but 2 of the hound's leaps are equal to 3 of the hare's: will the hound catch the hare, and if not, why?

8. A fish's head is 6 in. long; its tail is as long as the head and half of its body; and the body is as long as both head and tail: what is the length of the fish?

ANALYSIS.—*Since the tail is as long as the head and half the body, the tail is 6 in. $+ \frac{1}{2}$ the body; but,*

The body = the head and tail; therefore, the body = 6 in. $+$ 6 in. $+ \frac{1}{2}$ the body; that is,

The body = 12 in. $+ \frac{1}{2}$ the body; therefore, $\frac{1}{2}$ the body is 12 in., and the body is 24 in. long.

The tail = 6 in. $+ \frac{1}{2}$ of 24 in. = 18 in. Therefore, the whole length is 6 in. $+$ 24 in. $+$ 18 in. = 48 in.

9. A trout's head is 4 in. long; its tail is as long as its head and $\frac{1}{3}$ of its body; the body is as long as its head and tail: what is its length?

10. A has 10 ct.; C has as many as A, $+ \frac{2}{3}$ as many as B; B has as many as both A and C: how many ct. have B and C each?

11. A man bought a sheep, cow, and horse; the sheep cost $8; the cow, as much as the sheep, and ¼ as much as the horse; and the horse cost twice as much as both the sheep and cow: what did each cost?

12. The head of a fish weighs 8 lb.; the tail weighs 3 lb. more than the head and half the body; and the body weighs as much as both the head and tail: what is the weight of the fish?

13. B is pursuing A, who is some distance in advance; B goes 4 steps, while A goes 5, but 3 of A's steps = 2 of B's; B goes 30 ft. before overtaking A: how many ft. is A in advance of B? *Ans.* 5 ft.

LESSON XII.

1. A gentleman meeting some beggars, found that if he gave each of them 3 ct., he would have 12 ct. left, but if he gave each of them 5 ct., he would not have money enough by 8 ct.: how many beggars were there?

ANALYSIS.—*Each beggar will receive* 2 *ct. more when there are* 5 *ct. given to each, than when there are* 3 *ct.; but,*

Since the money to be distributed is 12 *ct. more than* 3 *ct. for each beggar, and* 5 *ct. for each beggar is* 8 *ct. more than the money, it will, therefore, require* 20 *ct. more to give* 5 *ct. to each, than to give* 3 *ct.; hence,*

As each beggar gets 2 *ct. more, it will take as many beggars to get* 20 *ct. more, as* 2 *ct. are contained times in* 20 *ct., which are* 10. *Ans.* 10 *beggars.*

2. A father wishes to distribute some peaches among his children; if he gives each of them 2 peaches, he will have 9 left; but if he gives each 4 peaches, he will have 3 left: how many children has he?

Explanation.—The difference between having 3 peaches left or 9 left, is 6.

3. Mary wishes to divide some cherries among her playmates; she finds that if she gives each of them 5, she will have 21 left; but if she gives each 8, she will have none left: what is the number of her playmates?

4. A lady wished to buy a certain number of yd. of silk for a dress; if she paid $1 a yd., she would have $5 left; but if she paid $1½ per yd., it would take all her money: how many yd. did she want?

5. To buy a certain number of oranges at 8 ct. each, requires 6 ct. more than all the money James has; but if he buys the same number of lemons, at 3 ct. each, he will have 29 ct. left: how much money has he?

6. There are two pieces of muslin, each containing the same number of yd.; to buy the first at 12½ ct. a yd., requires 40 ct. more than to buy the second at 10 ct. a yd.: how many yd. in each?

7. Five times a certain number is 16 more than 3 times the same number: what is the number?

ANALYSIS.—5 *times any number less* 3 *times the same number, is* 2 *times the number; therefore,* 2 *times the number is* 16.

8. Thomas's age is three times that of James, and the difference of their ages is 10 yr.: what is the age of each?

9. The age of A is 5 times the age of B; and the age of B is twice the age of C; A is 45 yr. older than C: what is the age of each?

10. A has ½ as much money as B; B has ⅓ as much as C; C has $15 more than A: how much money has each?

11. A farmer's sheep are in 3 fields; the second contains 4 times as many as the first; the third 3 times as many as the second, and 70 more than both the first and second: how many sheep are there in each field?

12. The age of A is ½ the age of B; twice the age of A is ⅓ the age of C; C is 20 yr. older than B: what is the age of each?

13. A father, who had as many sons as daughters, divided $18 among them, giving to each daughter $2, and to each son $1: how many children had he?

14. A man agreed to pay a laborer $2 for every da. he worked; and the laborer, for every da. he was idle, was to forfeit $1; at the expiration of 20 da., the laborer received $25: how many da. was he idle?

ANALYSIS.—*Had he worked every da. he would have received, at the expiration of the 20 da., $40; but as he received only $25, he lost $15 by being idle.*

Each da. he was idle he received $3 less than if he had worked, $2 wages, and the $1 forfeited; hence,

He was idle as many da. as $3 are contained times in $15, that is, 5 da.

15. James was hired for 30 da.; for every da. he worked, he was to receive 30 ct., and for every day he was idle, he was to pay 20 ct.; at the end of the time, he received $5: how many da. did he work?

16. When A and B entered school, the age of A was 3 times that of B; but in 5 yr., A's age was only twice B's: what were their ages at first?

ANALYSIS.—*Since the age of A is three times that of B, and when 5 yr. are added to each, the age of A is only twice that of B, therefore,*

Three times the age of B, (which is the age of A,) increased by 5 yr. = twice the age of B and 5 yr; that is, = twice the age of B, and twice 5 yr.; hence,

The age of B must be 5 yr.; and that of A, 15 yr.

17. Four yr. ago, the age of A was 3 times the age of B; but 4 yr. hence, it will be only twice his age: what were their ages 4 yr. ago?

18. B's age is twice A's; in 10 yr., A's will be ⅝ of B's: what are their ages?

19. A person has 2 watches, and a chain worth $10; the first watch and chain are worth half as much as the second; the chain and second watch, are worth 3 times as much as the first: what is the value of each?

ANALYSIS.—*Since the 1st watch and chain are together worth 1-half of the 2nd, the 2nd must be worth twice the 1st + twice the chain. But,*
The 2nd watch and chain = 3 times the first; or the 2nd watch = 3 times the 1st — the chain; hence,
Three times the 1st — the chain = 2 times the first + 2 times the chain; or the 1st = 3 times the chain, or $30; and the 2nd = $80.

20. A person has 2 horses, and a saddle worth $12: the 1st horse and saddle are worth ⅓ of the 2nd horse; but the 2nd horse and saddle are worth four times the 1st horse: what is the value of each horse?

21. If a herring and a half cost 2 pence and a half, how many can you buy for 9 pence?

LESSON XIII.

1. If 12 peaches are worth 84 apples, and 8 apples 24 plums, how many plums shall I give for 5 peaches?

2. If 1 ox is worth 8 sheep, and 3 oxen are worth 2 horses, what is the value of 1 horse, if a sheep is worth $5?

3. A walks 10 mi. in $1\frac{1}{4}$ da., and B 8 mi. in $1\frac{2}{3}$ da.: how far will B travel while A is traveling 20 mi.?

4. A bought a number of apples at 2 for 3 ct., and as many more at 2 for 5 ct.; he sold them at the rate of 3 for 7 ct.: how much per dozen did he gain?

5. C bought a number of eggs at 2 ct. each, and twice as many at 3 ct. each; he sold them at the rate of 3 for 10 ct.: how much per dozen did he gain?

2d Bk.   10

If he had sold them at the rate of 4 for 10 ct., how much per dozen would he have gained or lost?

6. Bought a number of pears at 2 for 1 ct., and as many more at 4 for 1 ct.; by selling 5 for 3 ct., I gained 18 ct.: how many pears did I buy?

7. A poulterer bought a number of ducks, at the rate of 6 for $1, and twice as many chickens, at the rate of 8 for $1; by selling 2 chickens and 1 duck for $½, he gained $2½: how many of each did he buy?

8. If 3 men can perform a piece of work in 4 da., working 10 hr. a da., in how many da. can 8 men perform the same job, working 6 hr. a da.?

9. Divide 32 peaches between Mary, James, and Lucy, giving Mary 2, and Lucy 3 more than James.

10. If 10 gal. of water per hr. run into a vessel containing 15 gal., and 17 gal. run out in 2 hr., how long will the vessel be in filling?

11. A can do a piece of work in $4\frac{1}{2}$ da., and A and B together in $2\frac{4}{7}$ da.: in what time can B do it alone?

12. A, B, and C, together, can do a piece of work in 5 da.; A and B, in 8 da.; and B and C, in 9 da.: in what time can each of them do it alone?

13. If 5 men *or* 7 women can do a piece of work in 35 da., in what time can 5 men *and* 7 women do it?

14. If 2 men and 4 women can do a piece of work in 28 da., in what time can 1 man and 1 woman do it, if a woman does $\frac{3}{4}$ of a man's work?

15. A man and his wife consume a sack of meal in 15 da.; after living together 6 da., the woman alone consumed the remainder in 30 da.: how long would a sack last either of them alone?

16. A man had 80 eggs, which he intended to sell as follows: 36 at 3 for 4 ct., 24 at 4 for 3 ct., and the rest at 10 for 17 ct.: but having mixed them, how must he sell them per dozen to get the intended price?

17. If $\frac{3}{4}$ of James's money be increased by $6, the sum will equal what Thomas has; both together have $34: how much has each?

18. If $5 be taken from the $\frac{2}{5}$ of A's money, the remainder will equal B's; both together have $51: how much has each?

19. The age of A is twice the age of B; and $\frac{3}{5}$ of B's age + 44 years, = $2\frac{1}{2}$ times the age of A: what is the age of each?

20. A has not $40; but if he had half as many more, and 2\frac{1}{2}$ besides, he would have $40: how many has he?

21. Two and a half times a number + $2\frac{1}{2}$ = 100: what is the number?

22. A farmer sold $\frac{3}{8}$ of his sheep, but soon afterward purchased $\frac{4}{9}$ as many as he had left, when he had 65 sheep: how many sheep had he at first?

23. John had $50 in silver and gold; $\frac{5}{8}$ of the silver, increased by $10, is equal to $1\frac{3}{4}$ times the gold: what amount has he of each?

24. One-half of A's money, diminished by $3, is equal to $\frac{1}{3}$ of B's, increased by $5, and both together they have $56: how much money has each?

25. A started from C the same time that B started from D; when they met, $\frac{3}{4}$ of the distance A traveled = $\frac{4}{5}$ of the distance B traveled; from C to D is 86 mi.: what was the distance each traveled?

26. Two-thirds of A's money = $\frac{1}{4}$ of B's, and $\frac{3}{4}$ of their difference is $15: how much money has each?

27. My watch and chain cost $\frac{9}{2}$ as much as my watch; 3 times the price of my chain + twice the price of my watch = $100: what did each cost?

28. Three towns, A, B, and C, are situated on the same road; the distance from A to B is 24 mi.; and $\frac{7}{8}$ of the distance from A to B = $\frac{3}{4}$ of the distance from B to C: how far is it from A to C?

29. A, B, and C, rent a pasture for $92; A puts in 4 horses for 2 mon., B 9 cows for 3 mon., and C 20 sheep for 5 mon.: what should each pay, if 2 horses eat as much as 3 cows, and 3 cows as much as 10 sheep?

30. John bought 5 melons for 5 ct., and James 3 melons for 3 ct.; they then joined Thomas, and each one ate an equal part of the melons; when Thomas left, he gave them 8 ct.: how should this be divided?

31. A person having 3 sons, A, B, and C, devised $\frac{2}{7}$ of his estate to A, $\frac{1}{3}$ to B, and the remainder to C; the difference of the legacies of A and C was $160: what amount did each receive?

32. The age of A is $\frac{5}{3}$ of the age of B; and the sum of their ages + half the age of B = twice the age of A — 2 yr.: what is the age of each?

33. A and B together can do a job in 16 da.; they work 4 da., when A leaves, and B finishes the work in 36 da. more: in how many da. can each do it?

34. Three persons, A, B, and C, are to share a certain sum of money, of which A's part is $12, which is $\frac{2}{7}$ of the sum of the shares of B and C; and $\frac{3}{8}$ of C's share is equal to $\frac{3}{10}$ of the sum of the shares of A and B: what are the shares of each?

35. If 9 men mow a field in 12 da., how many men can mow $\frac{1}{3}$ of it in $\frac{3}{4}$ of the time?

36. Two men formed a partnership for 1 yr.; the 1st put in $100, and the 2nd, $200: how much must the first put in at the end of 6 mon., to entitle him to 1-half of the profits?

37. A and B had 24 ct.; A said to B, "Give me 2 of your ct., and I shall have twice as many as you;" B replied, "Give me two of yours, and I shall have as many as you:" how many had each?

38. A gentleman being asked his age, replied: "The excess of $\frac{3}{5}$ of 50 above my age, is equal to the difference between my age and 10 yr.:" what was his age?

39. If I sell my eggs at 6 ct. a dozen, I will lose 12 ct.; but if I sell them at 10 ct., I will gain 18 ct.: what did they cost per dozen?

40. If I sell my sugar at a certain price per lb., I will lose $1, but if I increase the price 3 ct. per lb., I will gain 50 ct.: how many lb. have I?

41. If the labor of 1 man is equal to that of 2 women, and the labor of 1 woman is equal to that of 3 boys, how many men would it take to do in 1 da., what 12 boys are a wk. in doing?

42. If sugar worth $3\frac{1}{2}$ ct. a lb., be mixed in equal quantities with sugar worth $6\frac{1}{2}$ ct., what will $\frac{1}{2}$ a lb. of the mixture be worth, and how many lb. must be given for $1?

43. If $\frac{2}{3}$ of the gain $= \frac{4}{15}$ of the selling price, for how much will $3\frac{3}{4}$ yd. of cloth be sold, that cost $4 a yd.?

44. When sugar is worth 7 ct. a lb., a package was sold for 24 ct., gaining 3 ct.: for how much should a package weighing twice as much be sold, to gain 5 ct., when sugar costs 8 ct. a lb.?

SECTION XXVI.—PERCENTAGE.

LESSON I.—GAIN AND LOSS.

Explanation.—The terms *Percentage* and *Per cent.* mean a certain number of parts out of each hundred parts; that is, a certain number of *hundredths* of the sum considered.

One per cent. of any number is 1-hundredth of that number, that is, 1 of each 100 parts.

Thus, 1 per cent. of $100 is $1: 1 per cent. of $200 is $2, of $50, 50 cents: of $1, 1 cent.

1. What is 1 per cent. of $1? $2? $5?
2. What is 2 per cent. of $3? $4? $6?

3. What is 3 per cent. of $10? $20? $60?
4. What is 4 per cent. of $25? $45? $75?
5. What is 5 per cent. of $100? $300? $700?
6. What is 6 per cent. of $150? $250? $350?
7. What is 2½ per cent. of $100? $200? $500?

8. I bought a piece of cloth for $15, and in selling it gained 5 per cent. of the cost: what did I gain?

ANALYSIS.—*5 per cent. of 1 is five 1-hundredths; 5 per cent. of $1 is 5 ct.; of $15 it will be 15 times 5 ct., which are 75 ct. Ans. 75 ct.*

9. A grocer bought a bl. of sugar for $10, and in selling it gained 10 per cent. : how much did he gain?

10. A farmer having a flock of 40 sheep, lost 5 per cent. of them: how many had he left?

11. A flock of 50 sheep increases 10 per cent. in one year: how many are then in the flock?

12. A lady having $20, spent 10 per cent. for muslin, and 10 per cent. of the remainder for calico: how much did she pay for both?

One per cent. of anything is $\frac{1}{100}$ part of it; two per cent. is $\frac{2}{100} = \frac{1}{50}$; four per cent. $\frac{4}{100} = \frac{1}{25}$.

5	per cent. = $\frac{1}{20}$		20	per cent. = $\frac{1}{5}$
8⅓	per cent. = $\frac{1}{12}$		25	per cent. = $\frac{1}{4}$
10	per cent. = $\frac{1}{10}$		33⅓	per cent. = $\frac{1}{3}$
12½	per cent. = $\frac{1}{8}$		50	per cent. = $\frac{1}{2}$
16⅔	per cent. = $\frac{1}{6}$		75	per cent. = $\frac{3}{4}$

13. I paid 30 ct. per yd. for muslin: at what price must I sell it, to make 10 per cent.?

ANALYSIS.—*Ten per cent. being 10-hundredths, or 1-tenth, I must add to the first cost 1-tenth of itself: but,*

One-tenth of 30 ct. *is* 3 ct., *and* 30 ct. *added to* 3 ct. *are* 33 ct. Ans. 33 ct. per yd.

14. To make 12½ per cent. profit, what must muslin be sold at that cost 8 ct. per yd.? 25 ct.?

15. To make 8⅓ per cent. profit, what must sugar be sold for that cost 6 ct. per lb.? 12 ct.?

16. To make 25 per cent. profit, what must calico be sold for that cost 12 ct. per yd.? 16 ct.? 20 ct.? 35 ct.?

LESSON II.

1. A merchant bought cloth at $5 per yd., and sold it at $7 per yd.: what did he gain on a yd.? how much per cent.?

ANALYSIS.—*Since he bought at $5 and sold at $7 per yd., he gained $2 on every $5, that is,*

He gained $\frac{2}{5}$ of the first cost; $\frac{1}{5}$ of this is $\frac{20}{100}$, and $\frac{2}{5}$ are twice $\frac{20}{100}$ which are $\frac{40}{100}$, or 40 *per cent.*

2. James bought a melon for 4 ct., and sold it for 5 ct.: what per cent. did he gain?

3. An orange was bought for 5 ct., and sold for 4 ct.: what was the per cent. of loss?

4. Thomas bought a watch for $4, and sold it for $6: what per cent. did he gain?

5. Henry bought a horse for $15, and sold it for $24: what per cent. did he gain?

6. A keg of wine holding 5 gal., lost 6 qt. by leakage: what was the loss per cent.?

7. By selling citrons at 6 ct. each, John cleared $\frac{1}{5}$ of the first cost: what per cent. would he have cleared by selling them at 8 ct. each?

8. A merchant bought cloth at the rate of 6 yd. for $3, and sold it at the rate of 5 yd. for $4: what per cent. did he gain?

9. Henry sold melons at 8 ct. each, and lost $\frac{1}{5}$ of the first cost: what per cent. would he have lost by selling them at 3 for 25 ct.: what per cent would he have gained by selling them at 2 for 25 ct.?

10. James bought a lot of lemons, at the rate of 2 for 3 ct.; but finding them damaged, he sold them at the rate of 3 for 2 ct.: what per cent. did he lose?

11. Sold a watch for $12, and gained 20 per cent.: what was the first cost?

ANALYSIS.—20 *per cent. is* $\frac{20}{100}$ *or* $\frac{1}{5}$, *hence the gain was equal to* $\frac{1}{5}$ *of the cost; therefore,*

The watch sold for $\frac{5}{5}+\frac{1}{5}=\frac{6}{5}$ *of the cost:* 12 *then is* $\frac{6}{5}$ *of what?*

12. I sold a piece of cloth for $26, and gained 30 per cent.: what did the cloth cost me?

13. If there is a gain of 40 per cent. when muslin is sold at 14 ct. a yd., what is the cost price?

14. By selling a horse for $45, there was a gain of $12\frac{1}{2}$ per cent.: what did the horse cost?

15. Sold a horse for $45, and lost 10 per cent.: what was the cost?

16. Thomas sold a watch for $21, and gained 75 per cent.: what did he pay for it?

17. James sold 10 oranges for 40 ct., and gained $33\frac{1}{3}$ per cent.: how much did each orange cost?

18. When an article is sold at $\frac{5}{4}$ of its cost, what is the gain per cent.?

19. When an article is sold at $\frac{3}{5}$ of its cost, what is the loss per cent.? at $\frac{7}{8}$? at $\frac{9}{10}$? at $\frac{17}{20}$?

LESSON III.

1. When the gain is 20 per cent., what part of the cost is equal to the gain? when it is 75 per cent.?

PERCENTAGE. 157

2. When the gain is 100 per cent., what part of the cost is equal to the gain? when it is 150 per cent.?

3. When the loss is 25 per cent., what part of the cost is equal to the loss? when it is 35 per cent.?

4. What is the loss per cent. when the whole is lost? What is the gain per cent. when the gain is three times the cost?

5. When $\frac{1}{3}$ of the gain is equal to $\frac{1}{5}$ of the cost, what is the gain per cent.?

Explanation.—If 1-third of the gain is equal to 1-fifth of the cost, the whole gain is equal to 3-fifths of the cost, and 3-fifths are 60-hundredths or 60 per cent.

6. A sold a watch, so that $\frac{2}{3}$ of the gain was equal to $\frac{6}{25}$ of the cost: what did he gain per cent.?

7. When $\frac{5}{6}$ of the gain is equal to $\frac{15}{24}$ of the cost, what is the gain per cent.?

8. Sold a watch for $10, by which I gained 25 per cent.: what per cent. would I have gained by selling it for $12?

9. By selling muslin at 7 ct. per yd., there is a loss of $12\frac{1}{2}$ per cent.: what will be the loss per cent. by selling it at 6 ct. per yd.?

10. By selling my horse for $35, there was a loss of $16\frac{2}{3}$ per cent.: what would have been the gain per cent. by selling him for $63?

11. I bought a watch for $18, which was 20 per cent. more than its value: I sold it at 10 per cent. less than its value: what sum did I lose?

12. A sold B a watch for $60, and gained 20 per cent.: afterward B sold it and lost 20 per cent. on what it cost him: how much did B lose more than A gained?

13. A watchmaker sold 2 watches for $30 each: on one he gained 25 per cent., and on the other he lost 25 per cent.: how much did he gain or lose by the sale?

14. By selling 4 apples for 3 ct., a dealer gains 50 per cent.: what per cent. will he gain by selling them at the rate of 5 for 4 ct.?

15. Sold 5 lemons for 4 ct., and lost 20 per cent.: what per cent. will I lose by selling 6 for 5 ct.?

16. Two-thirds of 10 per cent. of 60, is $\frac{1}{2}$ of what per cent. of 40?

17. One-half of $\frac{3}{5}$ of 50 per cent. of 120, is 10 less than 20 per cent. of what?

LESSON IV.—INTEREST.

EXPLANATION.—*Interest* is money paid for the use of money.
The *Principal* is the sum of money which is loaned.
The *Amount* is the principal and interest added together.
The *Rate Per Cent.* is so many cents paid on each dollar.

1. If the interest of $1 at 6 per cent. for 1 yr., is 6 ct., what will be the interest of $10? of $12? of $15? of $20?

2. What is the interest of $2 for 3 yr., at 5 per cent.?

ANALYSIS.—*The interest of $1 for 1 yr. at 5 per cent., is 5 ct.; and for $2 the interest is twice as much as for $1, which is 2 times 5 ct., equal 10 ct.; and,*

For 3 yr. the interest is 3 times as much as for 1 yr., which is 3 times 10 ct., equal 30 ct. Ans. 30 ct.

3. Find the interest of $5 for 2 yr., at 6 per cent.
4. Find the interest of $8 for 5 yr., at 5 per cent.
5. Find the interest of $20 for 3 yr., at 8 per cent.
6. Find the interest of $25 for 6 yr., at 4 per cent.
7. Find the interest of $40 for 4 yr., at 5 per cent.
8. Find the interest of $50 for 3 yr., at 6 per cent.

PERCENTAGE.

9. Find the interest of $60 for 2 yr., at 7 per cent.
10. Find the interest of $75 for 3 yr., at 4 per cent.

LESSON V.

1. What the interest of $50 for 5 mon., at 6 per cent.?

ANALYSIS.—*For 5 mon. the interest will be 5 times as much as for 1 mon.; and for 1 mon., 1-twelfth as much as for a yr.*

2. Find the interest of $60 for 4 mon., at 5 per cent.
3. Find the interest of $80 for 7 mon., at ❋ per cent.
4. Find the interest of $40 for 9 mon., at 8 per cent.
5. Find the interest of $75 for 8 mon., at 9 per cent.

What is the interest
6. Of $120 for 6 mon. 15 da., at 5 per cent.?
7. Of $150 for 10 mon. 10 da., at 4 per cent.?
8. Of $45 for 11 mon. 23 da., at 8 per cent.?
9. Of $200 for 4 mon. 24 da., at 6 per cent.?
10. Of $480 for 9 mon. 18 da., at 5 per cent.?
11. Of $360 for 5 mon. 19 da., at 5 per cent.?
12. Of $144 for 8 mon. 25 da., at 4 per cent.?
13. Of $40 for 1 yr. 4 mon., at 6 per cent.?
14. Of $60 for 2 yr. 3 mon., at 5 per cent.?
15. Of $75 for 1 yr. 3 mon. 6 da., at 4 per cent.?

What is the amount
16. Of $25 for 3 yr., at 4 per cent.?
17. Of $40 for 2 yr., at 5 per cent.?
18. Of $55 for 3 yr., at 8 per cent.?
19. Of $30 for 1 yr. 4 mon., at 7 per cent.?

20. Of $50 for 2 yr. 3 mon. 6 da., at 6 per cent.?
21. Of $90 for 1 yr. 3 mon. 6 da., at 8 per cent.?

LESSON VI.

1. The interest of a certain principal for 2 yr., at 6 per cent., is $3: what is the principal?

ANALYSIS.—*The interest of $1 for 2 yr., at 6 per cent., is 12 ct.; and the principal must be as many times $1 as 12 ct. are contained times in $3. Ans. $25.*

2. The interest of a certain principal for 3 yr., at 4 per cent., is $6: what is the principal?

3. What principal at interest for 4 yr., at 5 per cent., will produce $12 interest?

4. What principal at interest for 5 yr., at 8 per cent., will produce $30 interest?

5. What principal at interest for 4 yr., at $7\frac{1}{2}$ per cent., will produce $42 interest?

6. What principal at interest for 2 yr. 6 mon., at 6 per cent., will produce $36 interest?

7. What principal at interest for 3 yr. 4 mon., at 6 per cent., will produce $70 interest?

8. A father wishes to place such a sum at interest at 5 per cent., as will produce for his son an annual income of $200: what sum must he invest?

LESSON VII.

1. What principal on interest for 2 yr., at 5 per cent., will amount to $55?

ANALYSIS.—*The amount of $1 for 2 yr. at 5 per cent., is $1.10, and it will require as many times $1 to amount to $55 as $1.10 is contained times in $55.*

PERCENTAGE.   161

What principal on interest,
2. At 6 per cent., for 3 yr., will amount to $236?
3. At 5 per cent., for 4 yr., will amount to $600?
4. At 10 per cent., for 5 yr., will amount to $375?
5. At 6 per cent., for 5 yr., will amount to $390?

6. The amount due on a note which has been on interest 3 yr. 4 mon., at 6 per cent., is $30: what is the face of the note?

7. Two-fifths of A's money on interest for 2 yr. 6 mon., at 8 per cent., is $60: what is his whole money?

LESSON VIII.

1. In what time, at 6 per cent., will $50 give $10 interest?

ANALYSIS.—*The interest of $50 for 1 yr., at 6 per cent., is $3; and it will require $50 as many yr., to give $10 interest, as $3 is contained times in $10, which is $3\frac{1}{3}$.*

In what time,
2. At 5 per cent., will $40 give $8 interest?
3. At 8 per cent., will $75 give $15 interest?
4. At 10 per cent., will $60 give $16 interest?
5. At 5 per cent., will $140 give $24 interest?
6. At 6 per cent., will $25 give $10 interest?

7. In what time, at 4 per cent., will any given principal double itself?

ANALYSIS.—*At 1 per cent., any given sum, as $100, will double itself in 100 yr.; and,*

At 4 per cent., it will double itself in $\frac{1}{4}$ of the time that it will at 1 per cent.: $\frac{1}{4}$ of 100 yr. is 25 yr. Ans. 25 yr.

8. In what time will any given principal double itself, at 2 per cent.? at 3 per cent.? at 5 per cent.? at 6 per cent.? at 7 per cent.? at 8? at 10? at 12?

9. In what time will any given principal treble itself, at 5 per cent.?

10. In what time will any given principal treble itself, at 8 per cent.? at 10 per cent.?

LESSON IX.

1. At what per cent. will $200, in 2 yr., give $24 interest?

ANALYSIS.—*At* 1 *per cent. for* 2 *yr.*, $200 *will give* $4 *interest; and,*

It will take as many times 1 *per cent. for* $200 *to give* $24 *interest, as* $4, *the interest of* $200 *at* 1 *per cent., is contained times in* $24.

At what per cent.,

2. Will $50 in 5 yr., give $20 interest?
3. Will $75 in 3 yr., give $11¼ interest?
4. Will $300 in 3 yr., give $63 interest?
5. Will $300 in 2 yr. 3 mon., give $54 interest?
6. Will $240 in 3 yr. 4 mon., give $56 interest?
7. Will $200 in 4 yr., amount to $240?
8. Will $150 in 3 yr. 8 mon., amount to $183?
9. Will any given principal double itself in 20 yr.?

ANALYSIS.—*Any principal, as* $100, *will double itself in* 1 *yr., at* 100 *per cent., and in* 20 *yr., at* $\frac{1}{20}$ *of* 100 *per cent., or* 5 *per cent.*

At what per cent.,

10. Will any given principal double itself in 12 yr.?

PERCENTAGE.

11. Will any given principal double itself in 10 yr.?

12. Will any given principal double itself in 8 yr.? in 5 yr.? in 4 yr.? in 2 yr.?

LESSON X.

1. What principal, at 5 per cent. for 4 yr., will amount to $72?

ANALYSIS.—$1, *at 5 per cent., for 4 yr., will amount to* $1.20; *and,*

The required principal will be as many times $1, *as the amount of* $1 *at the given rate for the given time, which is* $1.20, *is contained times in* $72.

2. What principal, at 6 per cent., for 5 yr., will amount to $520?

3. What principal, at 4 per cent., for 5 yr., will amount to $30?

4. What principal, at 10 per cent., for 5 yr., will amount to $750?

5. What principal, at 5 per cent., for 3 yr., will amount to $345?

6. What principal, at 6 per cent., for 4 yr., will amount to $496?

7. What is the present worth of $24, due 4 yr. hence, reckoning interest at 5 per cent.?

Explanation.—The *present worth* is that principal of which $24 is the amount. The *discount* is the interest on the present worth.

8. What is the present worth of $65, due 5 yr. hence, interest at 6 per cent.? what the discount?

9. What is the present worth of $55, due 5 yr. hence, interest at 5 per cent.? what the discount?

10. A owes $77, payable 6 yr. 8 mon. hence: what will he gain by paying it now, money worth 6 per cent.?

LESSON XI.

1. At 6 per cent., for 4 yr. 2 mon., what part of the principal is equal to the interest?

2. At 5 per cent., for 5 yr., what part of the amount is equal to the interest?

3. When the interest for 2 yr. $= \frac{1}{5}$ of the principal, what is the rate per cent.?

4. When the interest for 2 yr. 6 mon. $= \frac{1}{4}$ of the principal, what is the rate per cent.?

5. When the interest, at 10 per cent. $= \frac{3}{5}$ of the principal, what is the time?

6. When 3 times the yearly interest $= \frac{9}{25}$ of the principal, what is the rate per cent.?

7. When $\frac{1}{5}$ of the interest for 2 yr. $= \frac{4}{25}$ of the principal, what is the rate per cent.?

8. When $\frac{5}{8}$ of the interest for 3 yr. $= \frac{9}{80}$ of the principal, what is the rate per cent.?

9. The interest for 8 mon. is $\frac{1}{25}$ of the principal: what is the interest of $200 for 1 yr. 4 mon.?

10. If the interest for 1 yr. 4 mon., is $\frac{3}{25}$ of the principal, what the interest of $100 for 1 yr. 8 mon. 18 da.?

11. In what time will any principal at 5 per cent., give the same interest as in 4 yr., at 10 per cent.?

12. The interest of A's and B's money for $3\frac{1}{3}$ yr., at 5 per cent., is $40, and A's money is twice that of B's: what sum has each?

13. Twice A's money $=$ 3 times B's; and the interest at 7 per cent. for $1\frac{2}{5}$ yr., of what they both have, is $49: how much money has each?

14. One-half of A's money $= \frac{2}{3}$ of B's; and the interest of $\frac{3}{4}$ of A's and $\frac{1}{2}$ of B's money, at 4 per cent. for 2 yr. 3 mon., is $18: how much has each?

THE END.

www.ingramcontent.com/pod-product-compliance
Lightning Source LLC
Chambersburg PA
CBHW030304170426
43202CB00009B/860